101 QUESTIONS ABOUT SCIENCE

101 QUESTIONS ABOUT SCIENCE

by

Brian J Ford

BOOK CLUB ASSOCIATES LONDON

This edition published 1983 by
Book Club Associates

By arrangement with
Hamish Hamilton Ltd, Garden House
57–59 Long Acre London WC2E 9JZ

British Library Cataloguing in Publication Data

Ford, Brian J.
   101 questions about science.
   1. Science – Examinations, questions, etc
   I. Title
   507'.6   Q182

ISBN 0 241 10992 2

Printed in Great Britain by Billing & Son
Worcester.

CONTENTS

Dedicated to Tamsin, Timothy and Leigh
for their inspiration

INTRODUCTION

I was lying on the beach of a tropical island at dusk. For a week before I had been giving university lectures, but this was the time to wash those teeming ideas from the mind. The sea lapped, the birds fluted and warbled, and my companion turned to me. 'This is the first time we have been alone,' she crooned, and ran her tongue over her lips. 'There is something I have been *longing* to ask you.'

'Oh, really,' I said, ' and what's that exactly?'

She moved a little closer, and smiled under lustrous eyes. 'Well . . .' she began, and the eyes blinked slowly, 'just how big is the average shooting star? In terms of mean particle size, that is, of course.'

'Of course,' I concurred. 'They are about the size of sand grains. And the longer-lasting trails are caused by fragments about as big as a peanut. Shooting stars are surprisingly small: meteors is their correct term, the streak of light they cause is a meteor trail. Occasionally you have one large enough to reach the ground, and that's when they are called meteorites.'

'Thank you so much,' she said, and lay back on the sand. 'I was dying to know that.'

It is surprising how much people want to know about science – compared with how little we bother to teach at school. Being bombarded with questions like that, in the most unlikely situations, is an occupational hazard of working in several different areas of science at the same time.

Another teaser came up on a BBC programme, when a schoolgirl asked: 'Why do Indian elephants have smaller ears than African elephants?' Nobody could tell her the answer, and I was left with the question, so I said: 'I should think it is because Africa is a bigger place than India; the elephants would have room to be more spread out, so they would need larger ears to pick up distant calls from other elephants.'

As I made plain, the answer was only a joke . . . and I do hope she realised that. But in fact many of the questions that children pose raise interesting puzzles and sometimes they show how little science knows. Question-asking should be greatly encouraged in education, I do not doubt; and honest failure to supply a good answer would be a powerful stimulus for research in the future.

In London, as I picked succulent morsels from a delicious sea-food concoction (and ran over in my mind the Latin names of the identifiable items as they emerged), my lunch-time companion confided: 'You know, I think that a gin and tonic is actually more intoxicating than a straight gin.'

'It probably is,' I answered.

'Really? But how could it be more intoxicating when it has been diluted by the tonic water? It doesn't seem to make sense.'

As I told him the answer, he smiled. Not for the first time he said, 'You should write a book about those things.' This time, I did. Do not regard these answers as definitive works of scientific argument, for they are nothing so grand. They are personal answers to questions I have been asked in recent months. Think of them as gleanings from an inquisitive scientist – jottings from an Agony Uncle.

By all means write to the publishers if there is a favourite you might like to see in the next volume; a few questions arising from this book have already triggered some examples for the next one. Dip in and out as you wish; this is meant to be light reading, not a textbook. You will certainly have noticed the spread of metric units in recent years, and since they are one's professional currency, so to speak, you will not be surprised to meet them in the text. But I have kept a sprinkling of miles, feet and tons to illustrate some concepts in a manner that may be more familiar.

And the gin-and-tonic answer? That is on p. 89. You see – there was an explanation all the while.

HOW DOES THE NEW COMPACT DISC WORK?

The main difference between a compact disc (CD for short) and the familiar long-playing record is the way the sound is stored. Conventional discs have a wavy groove in the surface which is an exact reproduction – or as near to 'exact' as possible – of the sound waves which were originally recorded. As the stylus follows the waving groove, the vibrations are recreated and in this way the sound that went in, comes out. There are several disadvantages which LPs present. First, they are easily scratched. Secondly, they wear out. And thirdly, they are twelve inches in diameter. The new CD record is much smaller, in fact you could lay one in this space. They are almost unbreakable, and the sound is preserved in the form of tiny indentations in the recorded surface of the disc. A beam of laser light is shone onto the disc as it turns, and the little reflecting marks modify the laser beam in a pattern that corresponds to the original sound waves. This complex signal is electronically converted back into an electrical current that rises and falls as an exact copy of the changes in the sound waves you started with in the studio, and when they are fed into a loudspeaker a replica of the studio sound emerges.

The disc itself can be fingered, it can be dusty, it can allegedly even be scratched; so long as the laser beam can shine through the protective, transparent coating on the disc and be reflected back by the tiny digital signals in the recording itself, then the sound will emerge as crisp, as fresh and as unadulterated as it did when the disc was clean and new.

The system does have disadvantages. The discs are much dearer than an ordinary LP, and the players even more so. The fact has not been widely raised, but it is true that conventional recordings on vinyl can be done as short runs, which encourages individual and private record producers to make discs for a limited market, or for audition purposes. Many groups have made their first recordings on small record presses. The CD system does not look likely to have a 'home-made' version available for local producers, so that would limit the chance for newcomers to make discs of their own.

CD records are at present marketed in ghastly little plastic cases that make nonsense of the compactness of the disc (the cases are larger than they should be) and are also fragile. That makes nonsense of the durability of the disc inside. In short, there can be no real doubt that the new discs will attract a fanatical audience from the start – and they have, indeed, already begun to do so – which will increase when the unit costs fall as production goes up. On the other hand, LPs will certainly last for a very long while. Many of them contain recordings that may never be available in the CD format, and newcomers to the music scene will need to make LPs if they do not have access to the capital-intensive CD production centres.

I think they are marvellous. But they are not – *quite* – the answer to everyone's prayer. There is a lot of life in the long-player.

HOW MANY GROOVES ARE THERE ON A LONG PLAYING RECORD?

Two. One on each side of the disc.

WHO INVENTED THE LONG PLAYER?

The classical accounts tell of records at $33\frac{1}{3}$ revolutions per minute being invented by Peter Goldmark at the Columbia Broadcasting System and introduced in 1948. But long playing records date back to the early years of the twentieth century, where they were used to provide the sound for pioneering 'talkies'.

Although sound-on-film was invented by Eugene Lauste in 1904 (and patented by him two years later), throughout the 1920s sound was popularly provided for film shows by the use of pre-recorded discs. These were larger than normal, 16 inches in diameter; they played from the inside to the edge, rather than from the outside in as do modern records, and they were made of the same brittle shellac as the better known 78s.

But they rotated at $33\frac{1}{3}$ rpm, and set the standard for today's long-playing album a quarter of a century before vinyl discs appeared in the shops. Goldmark's real breakthrough was in the development of vinyl records with a far more even, 'smooth' texture than shellac, which enabled them to carry a far finer groove than the old 78 rpm discs. Goldmark was actually a colour television scientist at Columbia Broadcasting, but he was irritated by having to change records frequently whilst listening to music at home in the evenings, and asked if he might start looking into new methods of making commercial recordings

instead. Columbia agreed, and Goldmark – working alone, and sometimes at his home in spare time – took about three years to perfect the idea. The Radio Corporation of America tried to compete with the Columbia initiative by bringing out a record playing at 45 rpm, but this has since been restricted to popular music only. The fact that the singles market is almost entirely 45s, and the album business exclusively $33\frac{1}{3}$ rpm LPs, is largely chance. If all discs were made to the $33\frac{1}{3}$ standard it might be better, since it would keep all record-players to a single standard.

And to go back to the question about the number of grooves on a long playing record: though there is only one, of course, its spirals are packed into the playing area of the disc at around 250–300 per inch. That is the theory. In practice, some companies like to sell more plastic than recording, you might say; by artificially spacing out the rate of cut when the master disc is produced, they are able to sell an album with a remarkably short playing time in total. If you are lucky enough to find an album on which the duration of each track is given (6'30" meaning six and a half minutes, for example) then it would pay to add up the total playing time so that you know how many minutes you are buying. It can be far less than you imagine.

HOW DOES A TAPE RECORDER WORK?

When recording, the sound waves picked up by a microphone are converted into electrical impulses and are then 'strengthened' by passing through an amplifier. The impulses are then fed through the coils of the recording head. The electrical current passing through the turns of the coil produces its own magnetic field, and the head acts as a tiny electromagnet, the strength of the magnetism increasing and decreasing exactly in proportion to the vibrations of the sound waves. As the tape passes across the head at a constant speed, the magnetic coating on the tape responds to the changing electrical field produced by the recording head, and the tape becomes magnetised in a pattern that is an exact replica of the electrical impulses we began with. And they, remember, are produced in a sequence that perfectly mirrors the sound gathered by the microphone.

As long as the tape is kept safe, it will carry with it – captured in its magnetic coating – this recorded sequence of impulses. If the tape is heated the magnetic message can weaken, since any magnet which gets hot tends to lose its magnetism. If it is passed through a strong magnetic field then the recording can obviously be damaged (which is why you should not stand tapes on top of a speaker cabinet, for there is a strong magnet inside the speaker). If it is struck – by dropping – then even that is enough to rearrange the magnetic molecules in the coating and cause your recording to deteriorate. And many people will have noticed that the impulses can sometimes print through from one layer of tape to the next, giving a quiet, barely audible 'echo' of one passage in another (which is why very thin long-playing tapes are not popular).

When the tape is to be replayed, it is drawn across the playback head at exactly the same speed as when it was recording. The magnetic impulses in the tape coating are now used to generate a series of impulses in the playback head identical with those originally generated in the recording head. Once again, these weak signals are amplified until they are powerful enough to drive a pair of headphones or a loudspeaker unit, recreating the original sounds. Some losses of quality are inevitable each time you pass from one form of energy to another. As you move from sound to electrical impulse, from there to the tape store, from the tape to the record head, and then after amplification back to sound again, you encounter at least four major steps – at each of which deterioration sets in. In spite of this the modern tape recording (even those using cassettes) produces a surprisingly high sound quality.

WHERE DID TAPE RECORDINGS ORIGINATE?

Tape recordings originated as wire recordings. In 1893 as Danish trouble-shooter named Valdemar Poulsen was taken on by the telephone company in Copenhagen. In his spare time he realised that you could magnetise wire to varying degrees if you drew the wire past an electromagnet, through which you were feeding current from a microphone. It took him five years to perfect the idea and patent it, and in 1900 his so-called 'telegraphone' took the Grand Prix at the great Paris Exhibition. By 1903 his idea was in production by the American Telegraphone Company, but there were practical difficulties and quarrels amongst the management, and the company closed down. Even when the 1920s saw the development of efficient amplifiers (which overcame one of the Telegraphone's chief drawbacks), Poulsen's invention remained lying fallow. It was not until plastic tape was invented by an Austrian independent research scientist, Dr Pfleumer, that interest began to return. The Allgemeine Elektrizitäts-Gesellschaft decided to market these new tapes for use with their Magnetophon recorder, and during the Second World War improved tape recorders were widely used in Germany.

In the United States the device was ignored until 1937, when a company called Acoustic Consultants began to produce a recording machine which they named the 'Sound Mirror', and the Bell Telephone Company used a steel recorder called the 'Mirrorphone' for pre-recorded telephone weather forecasts. The Armour Research Foundation gave America its first successful tape, superseding the supplies from Germany that were then in use. This organisation had received much impetus from the work of a student engineer named Marvin Camras, who made a high efficiency wire recorder in his spare time. The recording speeds of 30 inches per second which were used at the end of the 1930s were halved, then halved again by Armour, giving the present-day familiar standard of $7\frac{1}{2}$ and $3\frac{3}{4}$ ips but still with high quality. This lower speed was halved again, to $1\frac{7}{8}$ ips, when the popular 'compact cassette' was introduced. So the strange-sounding speed used by the world's cassette recorders is a direct descendent of recorders built half a century ago.

WHY DOES WATER PUT OUT A FIRE?

It doesn't always do so, of course. Water produces acetylene in contact with sodium carbide, an industrial chemical; and acetylene is a dangerously explosive gas. So any attempt to squirt water onto a fire where sodium carbide was in store would make things worse, not better. Water makes the metal sodium burn, too, and because water is a conductor of electricity (except when extraordinarily pure) it is dangerous to use on an electrical fire. Water poured onto burning oil or petrol will do nothing to extinguish the flames, because oil floats on water. But the effect can be to cause the burning liquid to spread. In these ways, water can *encourage* burning!

Water is an excellent fire-extinguishing liquid for burning materials like wood, paper and buildings. Partly this is because water prevents the air from coming into contact with the burning materials, which puts them out, but that is only of secondary importance. The main reason why water is so effective is that it takes a great deal of energy to vapourise it (this is known as its latent heat of vapourisation). When a piece of paper is burning, the heat produced by the flames starts off the process by which the neighbouring paper breaks down, giving off inflammable gases, and so extending the burning process. If water is sprayed onto the paper, virtually all the heat energy is absorbed in an attempt to evaporate the water. There is so little 'spare' energy that the remaining paper never enters the burning phase, and so the fire goes out.

Merely pumping water into a burning structure is no good, then, unless it is directed at the right place. When the French transatlantic liner *Normandie* was accidentally set alight in New York harbour during conversion to a troop-carrier in 1943, for instance, the local fire services poured so much water into the ship from their hoses that she filled with water and sank. In a case like that, a blanket of gas from some kind of extinguishing system would have been far more sensible. Ironically, the French had initially sent the *Normandie* to New York so that she would be safer from enemy attack.

HOW DO FIRE EXTINGUISHERS WORK?

There are several kinds of extinguisher. Some of them rely on pumping water out as a spray or a jet that can be directed towards the burning focus. One popular type is filled with a weak solution of sodium bicarbonate. At the far end (i.e. away from the nozzle) is a glass bottle of strong sulphuric acid. To operate, the extinguisher is struck at this end, where the knob breaks the concealed bottle of acid. The effect of the acid and the alkaline solution is exactly like a super-charged dose of indigestion salts: the contents of the extinguisher foam up under pressure and squirt out. The water is the main extinguisher here, but the carbon dioxide bubbles produced in the process help to produce an invisible blanket of gas that keeps out fresh air and hence deprives the fire of oxygen.

Carbon dioxide will not support burning, and it is a gas that is heavier than air, so it tends to lie around on the ground for a while and helps smother a fire. Some small extinguishers contain nothing but pressurised carbon dioxide. Others contain high-pressure soda water, sometimes with a foaming compound which will blanket the fire with bubbles. Sometimes dry sodium bicarbonate is packed into a cylinder and released by a charge of pressurised carbon dioxide (or nitrogen gas). The powder blankets the fire, keeping out the air, and liberates carbon dioxide on contact with anything that is still hot.

The most useful of these – where there might be a hazard from bringing water into contact with electricity – must surely be the carbon dioxide pressurised extinguisher. Here the gas is contained in a strong cyclinder under a pressure of several atmospheres. When the pressure is released, the gas rapidly expands as it escapes through a 'snow

tube' and instantly cools to well below freezing point. It thus blows a torrent of icy, frost-laden air and carbon dioxide towards the fire, lowering the temperature and the oxygen levels at the same time and controlling the blaze.

Carbon tetrachloride is another non-inflammable liquid used to put out fires. This is a colourless liquid which produces a heavy vapour when heated. When fires are likely in machinery or electrical equipment this kind of extinguisher is most useful.

The idea of keeping out the air is one good way of controlling small fires, and this is why people are recommended to throw a blanket around a person with burning clothes and throw them to the ground, rolling them about at the same time. I wonder whether that might tend to force hot cloth against the victims' skin, making their burns even worse. Throwing a coat, a blanket or rug, even a carpet or towel, over their burning clothes would rapidly extinguish the flames without running that risk and my guess is that this would be a safer answer to the problem. Keeping a fire

extinguisher handy is the safest solution of all, of course.

HOW DO THERMOMETERS FUNCTION?

When anyone asks about thermometers, I always remember the advice of the Scottish laird who instructed his people always to buy a thermometer in summer. 'You get that much more mercury,' he explained. In a way he is right. You buy the same amount of mercury, weight-for-weight, but it does expand in volume during the summer when the temperature rises. What he forgot was that the mercury might just as well have been bought on the coldest of days . . . it would have expanded just the same when warmed up, of course.

A conventional mercury thermometer does not work just because mercury expands when it becomes warmer, but because mercury expands more than the glass in which it is held. If the glass expanded at the same rate as the mercury inside –

if they had the same coefficient of expansion, we would say – then no matter what the temperature, the thermometer reading would remain unaltered. We tend to forget that the glass also expands, even if we ignore that in practical terms.

Mercury solidifies at $-39°C$, so for measuring temperatures around that level or lower a thermometer using dyed alcohol is preferred; alcohol has a far lower freezing point ($-117°C$ for ethyl alcohol, C_2H_5OH; $-98°C$ for methyl alcohol, CH_3OH).

WHAT ABOUT THERMOMETERS THAT CONTAIN NO LIQUIDS?

This particular bit of equipment is less common, but it is used for dial-reading thermometers. Here the designer produces a strip made from two metals fixed back-to-back, sandwich fashion. The metals have differing coefficients of expansion, so that one of them expands more than the other. The strip is wound into a coil, at the end of which is a mechanism that moves the hand on the dial. As the temperature rises and falls, so does the needle. Devices like this are popular for desk-mounted instruments and thermometers that make up a set for mounting on a wooden block and hanging on the wall.

A second kind of 'dry' thermometer isn't quite as liquid-free as it looks. This is the modern digital liquid-crystal strip. Here the numbers corresponding to temperature appear to materialise out of nowhere. In fact they are present all the time, as transparent figures in a black sheet of plastic. Behind them is a layer of temperature-sensitive liquid crystal material which changes its colour at specified temperatures. As the temperature rises, the band of pale-coloured crystal seems to migrate along the strip, revealing the figures corresponding to that temperature as it does so. Thermometers of this kind are often seen permanently fixed to the sides of fish-tanks, or temporarily fixed to the fore-heads of sick children, where they provide an easy indication of what is going on inside.

WHAT IS A THERMOCOUPLE?

If two wires of different metals are fixed together and one is kept a constant temperature while the other is heated, then a small but measurable electric current is generated at the junction. These systems are known as thermocouples, and typical pairs of metals used to make them are copper and constantan (an alloy of nickel and copper), or copper and iron. Only a few milliamps are produced, however; the current can either be amplified electronically for recording on a chart by means of an ink pen moving across a sheet of graduated paper, or an array of similar thermocouples can be built, which is known as a thermopile and produces larger voltages. This kind of device may be connected to a digital read-out system, to display the temperature in lights on the top of a public building, or as part of a prominent clock on display. Thermocouples are valuable for recording high temperatures and have particular applications for furnaces – where an ordinary glass thermometer would be destroyed.

HOW DOES A WASHING MACHINE WORK?

Every time I have been shown the booklet for a washing machine, it is always set out to look far more complicated than it is. There are lists of complicated programmes, which leave many users baffled and bemused. In fact washing machines with an automatic system run through a series of set operations. The first programme typically takes the machine through the whole lot, starting with a pre-wash, then going through several washing and rinsing cycles during which a very high temperature is reached (nearly boiling, as a rule), and then finally rinsing and tumbling in clear water before spinning dry. It is easier to build up what I call a 'total vision' of the way the machine works if you translate the complex booklet into a continuing flow of processes.

Starting at the high numbers, the last of these

settings would be for spin dry only (say, No. 12). The previous setting, No. 11, would be tumble in clear water, followed by No. 12, spin. Moving back to No. 10, that would be the setting for rinse and drain – followed by clear tumble, then spin. Setting No. 9 might be the coolest of the washes, followed by processes 10–12; No. 8 a warm wash (because, being an earlier number, the water has longer to heat up); No. 7 a hand-hot wash; No. 6 hotter still; and so on back towards No. 1, which would be a pre-wash.

Once you have this system in mind it is easier to select the programme you want. Some booklets given out with machines seem to imply that you cannot use a machine for spinning alone, for instance; but the 'total vision' approach solves that difficulty when you are trying to understand your own machine.

Many automatic machines now have two separate sets of programmes, a high-fill for synthetics, and a low-fill for natural fibres like cotton and cotton blends. The low-fill allows the tumbling clothes more freedom to move, and so washes them more actively; but the cotton programmes usually have a longer spin, which gets out the extra water that natural fibres absorb. The synthetics are more easily given a permanent crease, so their high-fill programme has a more gentle spin at the end. It is important not to interchange the two, or you would end up with dripping wet cottons or, alternatively, creased synthetics.

There are two main types of machine, the tub machines (including paddle-wheel and agitator types) and the fully automatic or semi-automatic drum machines. The tubs are fed and emptied from the top, whilst the drums are usually opened from the front through a glass door. The tub machines are usually supplied with a water heater and have many automatic features, which allow you to set washing times and rinsing times, etc.

The drum machines contain a perforated stainless steel drum fitted with baffles that lift and drop the clothes as the drum rotates. They are often fully automatic, which makes it possible to set a complete programme; as explained above, the earlier the number of the dial, the more lengthy and (if washing) the hotter the process will be.

The water entering the machine is regulated by a valve in which there is a diaphragm that is sensitive to pressure. As soon as the desired level of water is reached, the pressure it exerts on the diaphragm makes it move inwards, switching a relay that closes the water inlet valve. At the same time a series of metal cams begins to turn slowly, allowing the various functions of the machine to take place in strict sequence.

WHAT DO I DO IF THE MACHINE GOES WRONG?

The top-loading machines are always said to have an advantage in that – if they suddenly stop in mid-wash – you can open the lid and take out the clothes, without waiting for the engineer.

There is a way of doing this with front-loading automatics, however. You should slide the machine away from the wall, making sure it is switched off at the main, and then lean it backwards until the water level is no longer showing through the glass door. Most machines only have an inch or so of water above the lower rim of the door port, so this is quite easy. Leaving the machine leaning against the wall, the door can be opened without any water spilling, the clothes removed and placed into a bowl underneath the opening, and the door fastened shut again before the machine is restored to its upright position.

If a machine jams in mid-wash, switch it off, rotate the entire programme control until you return to the same number and then switch on again. If that doesn't work, try the next number higher in sequence (if on No. 5, go round the dial once and set again on No. $5\frac{1}{2}$ or No. 6). If that fails, turn it round again – with the main switch in the OFF position once more – until you come to the empty-and-spin number at the end of the programme, for that may at least clear the machine for you.

At the very end of your trials, you can lean the machine backwards as described above, remove the clothes, and ring for the engineer. At least you can send the half-washed items to the launderette meanwhile, and that should surprise the engineer when he arrives.

WHY DID MY ACRYLIC JUMPERS LOSE THEIR SHAPE?

Acrylic is a valuable modern fibre, but it has a very important characteristic which few of the public know about and which causes no end of problems for users. Plastics which melt may pass through what is called a 'second order transition'. Without being technical, this is a critical temperature above which the normally resilient fibre softens and becomes much easier to deform permanently. For acrylic fibres, this second order transition point is at 70°C. If you wash or iron an acrylic garment above this temperature, then it will become baggy and – as it cools – it will have permanently lost its shape. I know so many people who are convinced that acrylics are baggy fabrics, or that they – the wearers, that is – must have mysteriously lost weight or gone a strange shape overnight. Not so. All that has happened is that the garment has had a hot wash, or a hot iron, and it will now be permanently baggy.

The answer is simple. Ensure that acrylics have warm washes (at any rate, below 60°C or 50°C to be safe) and that they never have an ironing above No. 1 (the lowest heat setting for a modern iron). They will thereby avoid crossing that impressive-sounding second order transition point, and will keep their shape and size for years; decades, in fact.

HOW SHOULD I MOP UP SPILLED TEA AND OTHER LIQUIDS?

There is a scientific answer for this popular problem. Whenever a pot of coffee or tea, or a glass of milk or beer is spilt onto the carpet, there seems to be a set sequence of responses:
a) blind panic;
b) frustration;
c) attempts to wipe up the liquid with wet cloths;
d) a profusion of hair-dryers and fan-heaters;
e) the end of a decent carpet.
The mistake is to use a wet cloth, which everyone in a straw poll has given me as their Number One Answer. It is a serious mistake, but is taught to us all at mother's knee. A cloth absorbs moisture through the capillary action which sucks the liquid into the spaces between the fibres of which the cloth is made. Now, if you try to use a wet cloth (or a damp one) to wipe up spilled liquids, you will find that it will not work. The reason is that the capillary pull of the cloth is greatly reduced, because the fine spaces between the fibres are already filled with water. That is why the cloth was wet to begin with.

When faced with a spilled liquid, there is only one answer – a bone-dry, washed cloth. A towel is ideal, as it has extra lengths of fibres to absorb the maximum amount of water when you dry yourself. If a pot of drink is spilt, grab a freshly washed towel which is springy, soft and bone-dry, and drop it onto the spillage. Press it well in, and carefully rub away at the wetted area. Carpets are usually made from threads that still contain natural oils, and so they take a long while to absorb liquids (for this reason, do not use an unwashed, new towel, for that will also contain natural oils and will not absorb the spillage fast enough).

The fresh towel, on the other hand, has an enormous mileage of empty capillary spaces through its structure, and it will draw the liquid away from the carpet and leave it almost as good as new. In our large family of six children, such spillages occur from time to time and even the worst examples – which included a tray of filled tea-cups deposited across the plain, unpatterned landing carpet – yielded easily to the treatment I propose. There is not even any need to try to dab the carpet clean afterwards, as virtually all the spillage disappears as if by magic inside the towel.

The sodden towel(s) should go for immediate washing, of course.

SO IS THAT THE BEST WAY TO DRY MY HAIR?

It is. Hair dryers are best left alone and reserved for blow-drying of hair that is merely damp. If you try to dry wet hair with a dryer, you run the risk of over-heating the strands (which can damage the hair) and by evaporating rinsing water you leave behind a deposit of whatever was dissolved in the water. Even in soft-water areas this can include a reasonable amount of detergent from the shampoo, which could affect the pH value (p. 77) whilst in hard-water conditions you can leave behind a deposit of mineral granules that make the hair rough.

A popular mistake is that people keep on one side a towel that is already damp and say 'use this for your hair'. That is bad advice. The towel relies on being bone-dry to exert its maximum 'pull' on the water it is attracting, and a damp towel will not dry hair well. Instead you should take the fluffiest and springiest towel you can find, and simply leave it in contact with your hair for a few minutes, moving it around, pressing it, dabbing it and so on; rubbing is unnecessary. In a short while the hair will be surprisingly dry, and the rinsing water – instead of being allowed to leave behind its dissolved components – will have been lifted out of your hair altogether.

WHERE DOES SNOW COME FROM?

Snow is not frozen rain. That is what we call hail. Snow forms when the temperature of moist air falls well below freezing. The air can then no longer contain its complement of water-vapour, and so the vapour begins to crystallise directly out as ice. The crystallisation usually begins on a dust particle which acts as a so-called 'nucleus', and the crystal grows gradually.

Because of the way the water molecules align as they pile up on the already-solidified surface, the snowflake grows out to form six arms. The way the ice is deposited depends on the conditions of the air at the time – how moist it is, how cold, etc –

and as two drifting snowflakes in a newly-forming snowcloud can never experience exactly identical forces at the same time, no two flakes can ever be identical.

The larger snowflakes that fall in heavy flurries are composed of many individual six-sided crystals. But fine snowfall, where individual crystals are precipitated, makes an entrancing sight under a microscope and is worth study even with a hand-lens like a magnifying glass.

In still air, the crystals can collect into larger flakes that may be several inches across. They are then shaped like up-turned cones, falling apex first to the ground. In air that is still for a considerable altitude, larger flakes can build up; examples as much as ten inches across have been recorded.

WHY ARE ICE SKATES THIN, BUT SKIS BROAD?

When pressure is applied to ice or snow crystals, melting tends to occur. This is how a snowball is made: the hands press together the ice crystals or snowflakes, and they undergo slight melting during the period of compression. When the pressure is released, the mass refreezes into a congealed whole. If the snow is too cold, it may be below the temperature at which this form of pressure melting will be induced. The day is then said to be 'too cold for making snowballs' – quite correctly. The answer is to press harder, and to squeeze the ball for longer than normal. You can still make a snowball even in very cold conditions with that in mind.

Now let us see how skis and skates operate. The skates have fine edges so that the weight of the skater is brought to bear on a small area of ice. The ice melts as the pressure of the blade is exerted on it, and the skate literally floats in a lubricating film of freshly-melted water, which instantly freezes once the skate has passed over. Here too, it can be so cold that the skater has to slow down and give the ice time to melt and act as a lubricant.

Skis work the 'other way up'. Here you have a flat surface to the ski, rather than to the ice as in the case of skating; and the sharp edges are the corners of the snow crystals over which the skis pass. Here too the concentration of pressure causes momentary melting, allowing the ski to flow across the snow. Without this effect, it would be as hard to skate on ice as it would be on glass; and ski-ing on snow would be no easier than ski-ing on a sand-dune.

WHERE DOES SNOW GO?

The obvious answer is: 'It melts'. But that is only a part of the answer. In snowy areas, far more snow sublimes than melts.

Subliming involves the passing of a solid substance into a vapour, without the normal prerequisite of passing through the liquid state first. One example is a block of camphor or naphthalene mothballs. They gradually sublime, never melting, but simply disappearing into the air as time passes. The deodorant candles that stand inside a slotted container act in the same manner.

Snow does this too. On a dry and bright day, the snow begins to evaporate back into the air: it does not melt unless it rises above freezing-point. Sometimes this produces interesting effects. It means, for instance, that in parts of the Antarctic there are great meteorite fields, left behind by what I assume to be subliming snow-fields. What seems to happen is that the snow gently moves along like a glacier creeping downhill, bringing the occasional meteorite with it. Eventually, at the end of its run, the snow lingers long enough to sublime and deposit its meteorite components. Over tens of thousands of years, these accumulate and they are now being recognised as important sources of untouched meteorites for scientific study.

A second example of sublimation at work is when clothes are hung on the line in winter. They will quickly freeze as stiff as a board, if the day is cold enough; but never fear, they will dry virtually as fast as if the day was milder. The ice will sublime away, and in a few hours the frozen

clothes can be brought in bone dry – without once having thawed out in the process.

Having said that, one word of warning. I pointed this out to an inquisitive neighbour in a hard winter, and she hung out a dripping sheet on her line. It was soon frozen solid. Unfortunately, it was a breezy day and, when the wind caught the end of the sheet, it struck the corner of an out-house – and snapped clean in half. The broken sheet did dry in the end, by sublimation. However I think the scientific interest of that fact was rather overshadowed by the unforeseen tragedy of that major fracture.

WHAT IS WATER?

Water is a strange substance. If you were a scientist from another world looking at earth through an analysing telescope, you would probably conclude – if you had any sense at all – that life on earth was impossible, and all because of water.

Water is composed of two gases, oxygen and hydrogen, in the ratio of two atoms of hydrogen to one of oxygen in each water molecule (the well-known formula for water being H_2O, in acknowledgement of the fact). But water ought not to be a liquid at all! The other chemical compounds of similar complexity are mostly gases (carbon dioxide, CO_2, ammonia, NH_3, sulphur dioxide, SO_2 and so forth). Water's molecular alignment is more like that of a solid than a liquid, and it is virtually unique in that the solid form (ice) floats on the liquid (water). Any other substance would be expected to do the opposite. Were this the case, the earth's oceans would be solid ice, with only a thin film of water on top, newly melted by the sun's rays during daylight. The amazing fact that ice actually floats means that aquatic creatures can be protected by a layer of ice, which insulates them against further heat loss. That is how you can find fish swimming contentedly beneath your feet if you walk on a frozen lake in deep winter.

If water obeyed the laws of science it would freeze from the bottom of the lake, working upwards. The fish would be stranded at the top of the water, and on that basis life could never have evolved to the levels of complexity we see today.

Water is also a very corrosive substance. It eats through solid steel in a matter of months. We call the phenomenon 'rusting', true; but it is still a powerful example of corrosion at work. You may not see many examples of water acting as a corrosive agent in nature, but that is only because it has reacted with most chemical materials in the early days of the world's history. Iron was soon oxidised to iron ore, for instance, and now we reduce it back to metallic iron for our ships, cars, washing machines and refrigerators (p. 80). But once the water gets at it, it soon begins to revert to the oxidised iron ore form once again, so to complete the return of the iron to its natural, 'corroded' state.

Pour water onto lime and the solid mass will hiss, expand, crack apart; mix water with concrete and the mass will heat up as it sets (too much concrete mixed at one time can get hot enough to destroy itself in the process). Add some water to sodium hydroxide (p. 77) and the solution will boil as it forms. Truly, water is not the ordinary, harmless, predictable substance we like to believe.

Water has a strangely elevated surface tension, which is why you can float a solid darning-needle on the surface of a wine-glass of water. It also has an unexpectedly high specific heat (that is an index of the amount of energy you have to put into a material to make it rise in temperature).

So it is only because water is an exception to many of the laws of science that it exists at all. We depend for our lives on a substance which ought not to behave as it does – in itself a sobering thought.

COULD OTHER LIFE SYSTEMS EXIST WITHOUT WATER?

They would have to. Earth is poised about 93 million miles from the sun, yet we have the three forms of water around us (the solid ice, the liquid water, and the gas water-vapour). If we were just a little further from the sun we would become too cool, and all our water would solidify; if we were a little nearer we would be too hot to survive. So the chances of finding another world with just the right balance in its orbit around another star would be minute, to say the least.

But you could evolve other systems, even though it takes a smattering of chemistry to explain exactly how to go about it. One parallel to water, H_2O, would be ammonia, which can be written H_3N. Whereas water is composed of hydrogen and oxygen, ammonia is composed of hydrogen and nitrogen:

$$H_2O = H^+ + (OH)^- \qquad \text{water}$$
$$H_3N = H^+ + (NH_2)^- \qquad \text{ammonia}$$

Ammonia and water are both ionising solvents, which is to say that they allow molecules to break down into positively and negatively charged ions. This is a vital part of the chemical machinery on which life depends. The two compounds form analogous substances, and we find that the chemical groups $-OH$ and $-NH_2$ in the formulae above can replace each other in a host of chemical compounds. Thus if you substitute the $-OH$ radical for the $-NH_2$ alternative you can convert acetic acid to acetamide; similarly the $-OH$ group in alcohol can be replaced by the $-NH_2$ to form ethylamide:

	acetic acid	ethyl alcohol
water type ($-OH$)	CH_3COOH	C_2H_5OH
ammonia type ($-NH_2$)	CH_3CONH_2 acetamide	$C_2H_5NH_2$ ethylamine

Liquid ammonia might exist in alternative solar systems where the temperatures are much lower than they are around our sun, and could arguably form the basis of some life-form even in the outer planets of our own solar system.

You can think up other alternatives for the hydrogen and oxygen system. We might cite the cyanide radical $-CN$. Cyanide seems nothing but a poison to us humans, but many earth organisms (notably bacteria) metabolise cyanide, and it could possibly have the makings of the basis for an alien life-form. There are other oxides of nitrogen, sulphur, or carbon which could stand alongside the oxide of hydrogen that we know as water. Hydrogen sulphide, H_2S, or hydrogen fluoride, NF, might be alternatives for H_2O.

There are in earth life three protein components – the amino acids methionine, cysteine, and cystine – where you find the sulphur atom in the place you might expect oxygen to be:

$$
\begin{array}{c}
NH_3^+ \\
| \\
H-C-C_2H_4-S-CH_3 \\
| \\
COO^-
\end{array} \qquad \text{methionine}
$$

We could even look for possible systems in which carbon, the core of every organic compound, was replaced by an alternative element. If we take the simplest of these molecules, methane, its CH_4 structure could be modelled by silicon, Si; tin, Sn; or even germanium, Ge:

$$
\begin{array}{cccc}
H & H & H & H \\
| & | & | & | \\
H-C-H & H-Si-H & H-Sn-H & H-Ge-H \\
| & | & | & | \\
H & H & H & H
\end{array}
$$

methane   silicon   tin   germanium

NATURAL SUBSTITUTION FOR CARBON GAS

One property of carbon which makes it important as the centre of a life-system like ours is that it can form long-chain molecules. For many years it was imagined that these alternative substitutes behaved differently to the carbon compounds, but since then we have discovered the silicones. They have exactly the kind of long-chain molecular structure that you could see as the beginning of an alien life-system on some distant planet:

$$
\begin{array}{cccc}
CH_3 & CH_3 & CH_3 & CH_3 \\
| & | & | & | \\
-O-Si-O-Si-O-Si-O-Si- \\
| & | & | & | \\
CH_3 & CH_3 & CH_3 & CH_3
\end{array}
$$

By altering the CH_3 around, this could be exchanged for a host of other side-groups and in that way one might even see the beginnings of a parellel to our own genetic code, DNA.

There is even the possibility of exchanging the carbon dioxide principle for that of silicon dioxide (from CO_2 to SiO_2). Silicon melts at very high temperatures indeed, indeed it is hard to melt at all with modern furnaces, and it is tempting to speculate that in some white-hot world there might even be a life-form where silicon stood in for carbon.

The problem is that none of these compounds exhibit the essential contrariness of water; they do not so flagrantly disobey the laws of nature. It is in the fact that it is an exception to many rules that our own precious, life-giving water has its importance to life on earth, and it may just be that this peculiarity makes water – and life on our planet – unique.

HOW DOES WATER GET TO THE TOPS OF TREES?

This is a question I am asked by schoolchildren, who are taught that the maximum height of a column of water that the air can support at sea-level is around 32 feet. If you try to suck water up from a greater depth than that then, no matter how powerful your pump, the column will rise to 32 feet and no further. So: how can trees 200 feet high draw up water from their roots to the topmost branches?

The trees seem to infringe a law of physics. The way they do it is by never breaking the fine columns of liquid that occur in each tiny vessel within their stems. As the young tree commences to grow, there is a continuous column of water inside each of these vessels. Most of the vessels are as fine as a human hair, on average, and as the tree grows in size the vessels progressively lengthen. So fine are the vessels that as the tree approaches the critical 32 feet level, the column of water remains intact in each vessel. The water molecules keep their cohesion, as it is called, and no matter how tall the tree, the tiny columns of water remain unbroken.

The tree is therefore like a gigantic wick, drawing water from the soil and allowing it to evaporate through the leaves. In this way the raw materials for growth are carried from the soil to the tree's leaf-factories. Waste materials are left behind in the leaves eventually, and when the leaves are shed these unwanted residues return to the earth. I think we should regard leaf-fall as the plant's main method of excretion. Do not imagine that evergreen trees are immune from this process, for they also drop their leaves (even though they do not do it all at the same season).

There are two intriguing aspects of the tall trees and how they lift their water above the limits set by physics. The first is the ticking or snapping sound that some tropical trees emit. This is now known to be due to the explosive breaking of the water column in occasional vessels inside the trunk, when drought is threatening and the roots cannot maintain the supply demanded by the pull of the leaves. When the lowered levels of water reach a critical value, the water column literally snaps – and the plainly audible 'ticking' sound is the result.

The second example concerns the tradition of hitting a walnut tree to encourage it to fruit. A traditional tale is that a walnut tree needs a good beating if it is to crop heavily (a walnut tree, a dog, a carpet and a wife have something in common; that is one pleasingly traditional version I was once told by a rather belligerent customer quaffing light refreshments in a bar). Nobody has found an answer to that, as far as I know, but I think the solution may lie in the water columns held up by cohesion inside the trunk. If the tree is hit, then the shock waves are just the kind of force that would 'snap' the tiny columns of water. That would signal to the tree that it was threatened with a poor water supply, since part of its conducting tissue was now effectively out of action, and this is just the stimulus that makes a tree fruit heavily. Root pruning has a similar effect. So I'd guess that perhaps the old tradition of beating the walnut tree is a kind of restriction of water intake that is, as it were, the lazy man's way of pruning its roots.

WOULD BEATING IMPROVE OTHER TREES?

Almost certainly, yes. But remember that the walnut is unusual in being a fruiting tree that is more than 32 feet tall! Most fruit-bearing species (apples, plums, pears, oranges) are half this, so for them it would not work.

People often complain to me that they manure a tree, feed it and water it well, and still it doesn't fruit; or they have a flowering climber which has been given every attention and it won't flower. But that is the problem. Plants flower (and fruit) in order to ensure their survival. If they are well provided for, then there is little stimulus to act in this way. Restrict growth by root pruning – or perhaps by hitting the trunk of a tall tree – and by cutting down on the feeding, and the plant will flower and fruit as a defensive mechanism, to make sure that (even if it is threatened) its progeny will survive.

If you want the best behaviour from a flowering or fruiting species, then, do not be too kind. The reproductive urge in plants is best thought of as a response to an environmental threat of some kind.

WHY DO SOME KINDS OF PLANT GROW TALL, AND OTHERS NOT?

The best example here must be the mosses, compared with the conifers. Neither of these great groups of plants have flowers, and it does seem odd that you can have gigantic conifers – including the tallest trees, the sequoias of California – and huge tree-ferns, with towering gingkos in our parks and gardens, whilst the mosses never manage to be more than a few inches tall.

The reason lies in the plants' life story and their sex lives. These plants feature 'Alternation of generations'. They have two phases in their life cycle. One of these, the gametophyte, produces gametes – sex cells – which are not very different from our own. There is a female ovum which is a large resting cell, and a large number of small spermatozoids which swim along just like our own sperm cells do.

From the fertilised egg-cell develops the second generation, the sporophyte. This (as the name implies) produces spores. When these generate they give rise to the gametophyte generation with its sex cells, and so the cycle is repeated. It so happens that the conifers, the gingkos and the ferns all have a well-developed sporophyte generation. They grow large and leafy, and often to great heights. The gametophyte generation is tiny. In the ferns it exists as a little scale-like plant

known as a prothallus, which grows in damp conditions usually hidden from sunlight. Here it is easy for the sperm cells to swim to their target and fuse with the egg-cell when they get there. In the larger trees like the conifers, the little prothallus never even leaves the cone, but grows and matures within the cone's protection. There is no chance for the sperm cells to be lost, or to dry up en route.

But the mosses went the other way in evolution. They developed the gametophyte generation as their leafy plant body, reducing the sporophyte to a small spore-producing parasite, as it were, that grows at the end of the gametophyte branches. This immediately poses problems. The spermatozoids have to swim to the ovule, and if they tried to do so across the vast spread of a 'tree-moss' they would soon dry up and be killed. For this reason the mosses have had to remain small, and they have been restricted to places where they can attract enough moisture for the sperm cells to be able to swim safely to their target. If it were not for this fact, then mosses could have grown as tall as any fir-tree and we would know them as mighty and spreading plants, instead of diminutive and humble marsh-dwelling species that few people even notice.

CAN ANIMALS FORETELL THE WEATHER?

Certainly not. I have heard of people telling that the moles have dug very deep in autumn, or that the rooks have built their nests extra high in the trees in springtime, and from this arguing that a cold winter, or a calm summer, are ahead.

This is nonsensical. What the creatures are responding to is some characteristic of the weather they are experiencing at the time. The moles are going deep to find food, because it is unduly cold at the time, or perhaps because the soil is dry and the water table lower than usual. The birds are nesting where they can in the trees, and if the weather is stormy at the time they will clearly build lower than if the air is still.

Now, it may be that a stormy summer is statistically more likely to follow a windy spring than a still one, or that a sharp winter often follows a cold (or dry) autumn. In that round-about way you could see some sense in the observation. But it is entirely unnecessary to invest the creatures of nature with such metaphysical attributes. They are reacting to what they experience, and not to the future at all.

DOES THAT APPLY TO PLANTS?

Yes. Plants produce bounteous crops if the summer threatened them a little. If the weather was perfect for growth they will not need to produce fruit in abundance; if the climate was too severe then they will not have been able to. But a slight threat is reflected in their production of fruit, and it is not the other way round.

DO WE RELY ON TREES FOR OUR OXYGEN SUPPLY?

No, we do not. This is a popular view, and it is regularly advanced by the world's ecologists as an argument in favour of forest conservation, but it is a view that is faulty. Some impressive figures have been put forward for the volume of oxygen liberated by a square mile of dense tropical forest, like the Brazilian Matto Grosso vegetation. But this does not take account of the amount of oxygen taken in by the other processes that go on.

The oxygen is liberated during the hours of sunlight as the plants in the forest are growing. They take in simple materials such as carbon dioxide and water, and liberate oxygen to the air as they synthesise carbohydrates and other complex molecules. This process of light-building – photosynthesis – is the bedrock on which all earth life depends. Every animal feeds either on plant life, or else on animals that themselves fed on plants. In every case the original energy source turns out to have been a plant, capturing sunlight through the chlorophyll of the leaves in the production of the world's food supplies. The types of living organism that derive their energy from such strange sources as sulphur or iron compounds do not contribute to the global energy supply, so we can eliminate these insignificant exceptions from our discussion. (I mention this to discourage the youthful scientist who will point to a drawing of a *Beggiatoa* in an encyclopaedia and write in to say that it derives its energy solely from hydrogen sulphide, H_2S, and not the sun, nor any product of solar energy. Even that would be wrong, however, since this organism has recently been shown to be able to use other foodstuffs and does not rely exclusively on H_2S at all; but we digress.)

However, the outpouring of oxygen during photosynthesis is counterbalanced by two other factors: first, the taking in of oxygen as the plant undergoes its normal life processes during darkness – for plants consume oxygen during normal metabolism, just as we do – and second, the consumption of further oxygen when the dead remains of the vegetation decay and are returned to the soil. The equation for production of complex molecules and the giving out of oxygen is counterbalanced by an opposite trend for the taking in of oxygen during the breakdown of the molecules:

$$6CO_2 + 6H_2O \rightleftharpoons C_6H_{12}O_6 + 6O_2$$

$$\underset{\text{dioxide}}{\text{Carbon}} + \text{water} \rightleftharpoons \text{glucose} + \text{oxygen}$$

When we move forward (i.e. the top equation), energy comes in from the sun. Moving from right to left (lower arrow) oxygen is consumed but energy is released to the environment. This energy output from decaying materials explains why it is that a compost heap will become warm, and may become dangerously hot.

In the example above, I have shown how carbon dioxide and water can be made by the seemingly miraculous processes of photosynthesis into glucose and oxygen – a process that science cannot imitate. I have chosen glucose as a simple example; all the other complex materials right up to starch and cellulose are made by an elaboration of the same process. But if the plant product decomposes, then it is obvious that the

same amount of oxygen is going to be taken in as was originally given out.

The example of the Brazilian forests shows this clearly. There is hardly any plant matter left behind in the soil, and the fertile top-soil is surprisingly thin. This means that almost all the plant material built up by the present generation of plants is doomed to break down in time. The forests show a continuous recycling between minerals and plants, which goes on over millennia. So, although we can measure large amounts of oxygen produced by trees during the day, their net contribution of this life-giving gas to the atmosphere is minute since it is consumed again at night. We have to conclude that forests do little to give us the oxygen we breathe.

That does not mean we should decimate the world's tree populations, however. They provide cover for life-forms on which we depend, and stabilise the land surface. But these economic reasons – which are so typical of the materialistic way in which urban populations are trained to think – are not the most important issues to my mind. The real reason why we should preserve the forests is because they are beautiful, they are impressive, and they are natural. They have existed long before mankind, and will continue to flourish after we have disappeared no matter what havoc we wreak. We should preserve forest life out of respect for our environment, and out of a desire to show that we can harmonise with the world we happen to inhabit.

So – where does the oxygen come from? The overwhelming bulk of it has been produced by single-celled algae floating in the world's seas. These organisms are actually vital to our survival. If all the oxygen produced by trees (about which we hear so much) were to vanish, it would be hard to notice the difference. But if the oxygen from the microbes (of whom we hear much less) vanished overnight, so would we.

WHAT UNEXPLOITED FORMS OF NATURAL ENERGY ARE THERE?

Almost everything we see around us could provide energy. Waste matter is a rich energy source, which could be harnessed either through crude burning in a furnace to raise steam, or processed by biotechnology to provide useful raw materials for our use. The earth itself is a vast reserve of heat energy. If you envisage the world the size of an orange, the crust on which we live is no thicker than a sheet of paper and all the rest is heat. I know how impressive the towering mountains and plummeting ocean depths seem to us mere mortals, but the world is proportionately as smooth as a billiard ball, for all its apparent surface imperfections; and since we spend all our lives on the fine skin around a sphere of red-hot mineral it seems surprising that we have had any form of energy crisis at all.

Then there are wave energy from the seas, wind power, tidal power and many other forms of environmental effect that we might harness. In fact there is no shortage of energy at all. Our problems arise because we have taken to burning precious chemical reserves – like natural gas and oil – which in my view should be used to construct valuable items like plastic water-pipes that cannot corrode, plastic-insulated electrical cables that cannot perish, drugs and a host of important chemicals for the developing world and the industrialised nations. This is no more intelligent than burning banknotes to keep warm for a night or two. So there is certainly a shortage of petrochemicals, occasioned largely by a perhaps understandable lack of Arabian goodwill;

'. . . waste matter is a rich energy source . . .'

but there is no shortage of energy. Every alternative source has been investigated in the past, and as soon as there is some tiresome financial incentive then attention will focus on one or other possibility and from then on we will look back to the present oil-burning era as one of corporate short-sightedness.

Take, as one example, hydroelectricity. Water power has been used for thousands of years, and water-wheels powered by a small mountain stream dropping a few feet can drive power-saws, drop-forge hammers, grinding mills and a host of other machines that ought still to be in use. Electrical power generation is a modern refinement of this time-honoured idea, and in recent years the Ossberger turbine (which has adjustable vanes that can match differing rates of flow) has been introduced for sites where the fluctuating flow of natural, seasonal stream would otherwise have been hard to handle. Small hydroelectric plants using this kind of generator could be attached to streams and rivers in innumerable sites. The United States has already encouraged the construction of small-scale generators through the National Energy Act, which now requires regional electricity supply agencies to purchase extra electricity from anyone with a small generator that can be wired up to the supply grid. It is calculated that there are 1,400 river dams in the United States, and a recent survey has suggested that there are 65,000–70,000 further dam sites which might be suitable for small-scale installations.

The introduction of local schemes in China shows how far this can go in practice; in the last $1\frac{1}{2}$ decades approximately 90,000 small hydroelectric units have been installed, and of course small units harmonise well into the surroundings, and do little damage to the ecological balance of the area.

Another means of controlling environmental impact is to site installations in unpopulated areas. One such area is the Greenland ice-cap. Though we think of it as an icy wasteland, every summer huge amounts of the ice thaw and pour in seasonal torrents to the lowlands. A survey by the Institute for Training and Research of the United Nations proposes that this melted ice could be used to generate electricity. Their proposal would involve the construction of a system of channels and aqueducts to direct the water flow into hydroelectric generators. Nobody knows for sure how much water actually melts each year, and it is certain that the total varies from season to season. But there is calculated to be roughly 20 times as much fresh water held in the ice cap as exists in all the world's lakes and rivers put together.

The fate of the electricity would depend on the nature of demand. It could be transmitted under the oceans via giant cables on the sea-bed, and fed directly into national electricity supply grids. Alternatively it could be used on site for hydrolysis of sea-water to produce oxygen and hydrogen gas, which could then be pressurised into tanks and shipped as a fuel to replace oil.

So far hydroelectricity accounts for a quarter of the global electricity supply. But there are some giant plans on the drawing-board that would extend this. The Itiapu Dam in South America is planned to produce 12,600 megawatts of power (which is more than a dozen nuclear stations combined). In China there is a scheme for the Yangtze River which would double that, with a planned capacity of 25,000 mW, whilst a proposal for the Amazon Basin would reach the enormous output of 66,000 megawatts.

So far the areas with the greatest potential for hydroelectricity seem to have been slowest in exploiting it. Top of this table is Asia, with nearly one-third of the world's potential hydroelectric power potential, of which it has harnessed a mere 9%. South America claims one-fifth of the global potential, but has tapped only 8%. North America boasts one-sixth of the potential total, and has so far harnessed 36% of it; whilst Europe has a mere one-tenth share, but has exploited some 60% of that to date.

This examination of some of the possibilities for hydropower is not to be taken as endorsement of the most hare-brained schemes, for large scale development has a serious effect on environmental factors. Flooding of land can uproot long-established communities or lose productive farmland. The lakes that result can lead to an increase in water-borne diseases, and the change of water flow can damage the downstream river. So it is certainly better to aim at non-invasive methods of generating electricity.

It is not even necessary to build dams at all, in many instances. The river flow itself can produce a sizeable contribution to a region's electricity demand. The National Research Council in Canada, for instance, has developed a small generator that can be anchored to the river-bed and left there to generate power without any significant effect on the surroundings. Already they claim that there is a simple rule of thumb, to give planners an idea of where the scheme could be introduced. Anywhere there is a river five feet deep, flowing at five feet per second, you could install a generator with a capacity of 5 kW. Plans are now afoot to design rigs that could eventually be produced in sizes up to 120 ft diameter, and the projected figures for the output from these submerged plants show that the combined output of a few of the rivers in British Columbia could provide 38,000 mW for the consumer.

So there is much that could be done to provide us with power, without the need to erect nuclear power stations or burn fossil fuels irresponsibly. The example of hydroelectricity is just one of the alternatives we have available, yet it shows what might be done by harnessing the earth's abundant, natural reserves of power.

And if at the same time we began to be less profligate in our waste of energy – fewer gas-guzzling cars used needlessly, lower temperatures inside centrally-heated buildings, less wasteful illumination – then we could ensure the safety of the present world at the same time as guaranteeing ourselves the kind of future most people want.

HOW MUCH ENERGY DOES A HUMAN BEING CONSUME?

We run on about as much energy as a 100 watt light bulb. Saving the waste heat produced by human bodies in offices and homes is one valuable way of conserving natural energy, of course; but

'are you AC or DC?'

'. . . we run on about as much energy as a 100 watt light bulb . . .'

at the same time the economy with which we use our own energy reserves shows how remarkably efficient metabolic energy can be. I have no idea how much energy you would require to power artificial devices that could carry out such human tasks as walking, carrying, undoing bottles and cans and so on – but it would take a little more than 100 W. And of course there is no machine that could approach the human ability to look, listen, feel, experience and think.

Most of the energy in the body is used by the spleen and the liver combined and it is in the liver that the delicate chemistry that regulates the body's metabolism is controlled. The oxidation of stored sugars by the liver cells produces the heat which keeps us at a minimum 37°C. Over a quarter of the energy of a resting human goes to these organs. Just about a quarter goes to the muscles – including the beating heart – and one-fifth is consumed by the brain. The amount of energy used by the brain cells does not alter if you suddenly start thinking hard.

ARE FAT PEOPLE GREEDY?

Not necessarily. There seems to be a kind of thermostat in the body that dictates how much fat you should store. This is an unpopular view amongst dieticians, I might add; indeed I heard one most eminent nutritionist recently tell a meeting that you could predetermine the physique of anyone by feeding them the right diet in childhood. But although this approach is popular, I am convinced it is wrong. Most young boys will eat their fathers out of house and home, given half a chance; yet they gain weight – or stay more or less stable – because of the hormonal control of their bodies. At puberty the most underweight and feeble child can develop into a muscled and heavy adult without any great change in the amount of food they eat. Conversely, there are many adults who eat far more than the average person and yet who retain a slender figure.

One example is the actress from the international television series *The Onedin Line*, Jessica Benton, who played the part of Lady Elizabeth

Fogarty in the programme for ten years. She is 5 ft 3½ in, 7 st 8 lb, and has a featherweight appearance. Yet she eats a formidable diet that gives her twice the calorific intake she ought to require. Her typical daily diet runs as follows:

BREAKFAST: full English breakfast with bacon and egg, toast and sweet tea . . . . . .900 Calories
SNACK: chocolate and biscuit bar, nuts, fruit and coffee . . . . .550 Calories
LUNCH: Starter, mixed grill, creme caramel for dessert with coffee . . . . .2,000 Calories
AFTERNOON SNACK: Sandwich, chocolate and biscuit bar, tea . . . . .600 Calories
TEA-TIME: Convenience snack (perhaps fish fingers with French fried potatoes) . .350 Calories
SUPPER: Vegetarian meal, with husband; vegetables au gratin and potatoes . . . . .350 Calories
LATE SNACK: Peanuts, fruit . . . . .200 Calories
ALCOHOL: One glass of wine with lunch or supper
100 Calories
TOTAL: . . . . .5050 Calories

The average intake for a healthily active woman is around half that daily total. Yet there are other individuals (and it would be invidious to cite names here) who are overweight and yet who carefully watch their dietary intake. So the first conclusion is that you do *not* necessarily become fat by overeating.

On the other hand, you can most certainly lose weight by cutting down your intake. It may be (as I am personally convinced) that your target weight is preordained by your body's chemistry, and if that is the case then no amount of over-eating will make you put on extra weight. But if the target weight is too high to suit your wishes, then obviously you can reduce your weight by restricting what you take in.

There is no virtue in trying to suggest here which diet is the best; there are many plans and they suit different people in different ways. Many individuals need a permanent reminder of the discipline of dieting, so for them a set regime of fixed meals is by far the best. Other people have a set diet they adopt, which they have decided upon for themselves (my wife always goes to a diet of three open sandwiches a day, for instance). My personal belief is that it helps for many people to

feel that their normal habits are not being restricted too much, so that they lose any feeling of being 'penalised' for their trouble. In this case I prefer to point out how easily you can adapt your normal daily diet and still lose weight. If you like a fried egg and fried bread, for instance, then have the egg poached and with half a slice of toast instead. If you like fatty steaks and french fries, then have lean grilled steaks and a small helping of potatoes. Have a small helping of porridge rather than a bowlful, and small cups of unsweetened tea without milk instead of large mugs with cream . . . this approach enables you to keep some normality to a diet, and yet the intake per day expressed in Calories can be below 1000.

The only advice I would give is that you look carefully at so-called dietary foods made by commercial companies. A slice of crispbread made specially for dieters is not likely to contain a single calorie less than a slice of ordinary crispbread. And half a slice of ordinary bread from a regular loaf may contain no more either. Several surveys in recent years have shown that the diet-aid preparations sometimes rate far higher in the energy stakes than conventional foodstuffs, so be warned.

Finally, one small mystery!

Many people have asked me why they seem to gain weight just a little when they have been on a diet for a few days. Now, I do not know how widespread this phenomenon is. Neither do I know of any explanation for the phenomenon that has been put forward in the past. But, if it makes sufferers from this complaint any happier, it occurs to me that there might be one practical answer. If reserves of fat are consumed during the early phases of dieting, then it may be that the fat cells remain in position – but they replace their stored fat reserves with body water instead. Now, water is denser than fats and oils. So if part of the space previously occupied with fat was now taken up by water instead, the same volume of tissue would actually weigh a little more. I do not know of scientific evidence that this takes place normally, but you may at least be satisfied to know that there is one possible answer to an otherwise inexplicable phenomenon.

MUST I CHANGE MY DIET TO AVOID BLOOD PRESSURE, ATHEROSCLEROSIS OR A HEART ATTACK?

Nobody knows. It has long been assumed that a lot of fatty deposits in the blood vessels are related to an excess of fatty materials in the diet, but this is not necessarily so. The most recent dietary constituent for which there is some sound-looking evidence is dietary fibre. This material, usually supplied as bran, is known to keep food moving healthily through the intestinal tract (and that – as well as aiding digestion – helps reduce the incidence of cancer), and it does now seem that in some unknown manner it also acts against the tendency to develop heart disease.

But there are even objections to this. I am inclined to argue that the so-called benefits could also arise because of the mental change, or the alteration of life-style, that people undergo when they decide to watch what they are eating. This in itself may produce effects, and it may be that we do not yet know how to tell the one set of effects from the other. It may even be that the change of life-style is actually the most important effect of all.

One model you can apply to human behaviour is what I call the 'biological imperative'. In this concept, we expect to find that inborn mechanisms take over and direct our destiny when we act in a way that would otherwise threaten our kind. I find this a better explanation of heart attacks and kindred diseases than the dietary alternative. According to this view, the disease is a response that helps to eliminate from society people whose behaviour patterns are abnormal or harmful. What examples could one cite?

Let me take just two: one from the human world, and the other from the experimental animal situation in which you would not expect to find disease of this sort. The human example concerns the descendants of the Italian immigrants who lived in the town of Roseto, Pennsylvania. Of this town's 1,630 inhabitants, over 95% had descended from the same small Italian community. They were found to eat just as much of the greasy and fattening foods that most of the authorities associate with arterial disease, consuming large amounts of ham, frying their food in plenty of lard, enjoying all the gravy and dripping they could. They ate far more than we would think was good for them, and drank plenty of wine.

Yet over a six-year study, not one of the group aged under 47 died of a coronary. Among the older men, the death rate from heart attacks was half that of the neighbouring townships. How could this be, when they ate such a rich and unhealthy-looking diet? The obvious answer to spring to mind is a genetic one. But this can be disproved, for in the people who went from Roseto to live in the big cities, and who changed to the conventional American style of living, the death-rate from heart attacks was as high as you would expect.

The more likely explanation lay in their own life-style. The community in Roseto was carefree, caring, trusting and ebullient. Their family structure was mutually supportive, and in this boisterous community there was a virtual absence of crime. So there is one human example of the effects of life-style on a community that – apparently – does not eat a healthy diet. The comparison with the fate of their members who ended up in the big cities, and who became equally liable to heart disease as the rest of the urban community, provides confirmatory evidence in support of the contention.

But now let us look at the other side of the coin. What happens if we take a group that are ordinarily unaffected by such diseases, and introduce frustrations into the system? This was done some years ago in a Soviet laboratory. They established a community of chimpanzees in a large compound, where one male was enclosed with several females. In a short time they established a community, a social order, and harmony was established as the male and his harem lived out their lives in relaxation and peace.

Then the experimenters stepped in, by removing the male leader of the pack and placing him into a cage where he could see his group, but not get at them. An attractive young male was introduced into the community, and in turn he took control of the group and became the 'lord of the harem'. The effect on the dispossessed male was profound, for he would have outbursts of rage and fury alternating with periods of sitting disconsolately in the corner of his cage. The outcome was always the same – within three months the imprisoned male chimpanzee was dead of high blood pressure and atherosclerosis.

So it may be that the heart attack, and its related diseases, are in part a sanction against a harmful life-style. If that is the case, then it is a return to a harmonious and satisfying life-style that will do more to improve our community health than any amount of alteration in our diet.

'. . . suffering from high blood pressure and atherosclerosis . . .'

Remember, the fact that the overwhelming majority of people stay the right shape no matter what they eat, shows that it is not overeating that is the root cause. If it was, then the svelte Miss Benton whom we left on p. 20 would be as round as a barrel. The tendency to exercise more, and to eat Calorie-reduced foods (like the great American craze for 'lite' products and the 'lean cuisine') is obviously going to help. But it may be that it is the adoption of the new approach to living – which in turn makes people want to eat carefully – that holds the real, ultimate solution.

WHAT IS A CALORIE, ANYWAY?

This is not as easy as it seems, in fact a precise answer would involve a lot of facts. Whenever I say that, people reply, 'Oh go on, tell me anyway,' so here goes. To begin with, the Calorie we use in food studies is not the same thing as a calorie (which may explain where the capital 'C' came from the passages on the previous five pages). The original calorie was defined as the amount of heat energy you need to raise one gram of water through one degree Celsius. This is a small amount of heat energy, and so the large calorie, also known as the Calorie (with the capital letter) and as the kilocalorie, was introduced instead. This was 1000 calories, and it became the standard measurement of the energy value of foodstuffs.

However, there are other calories in existence, such as the International Table calorie, used for engineering steam tables. This is defined as $\frac{6}{860}$ international watt-hour, or 0.00116298 absolute watt-hours.

As measuring accuracy increased over the years, it became apparent that the original definition of a calorie was not sufficiently precise, and it was redefined as 4.1833 international joules in 1930. In 1948 it was revised again, so that 1 calorie = 4.1840 absolute joules, and it has remained that value to this day. But not quite. Modern conversion tables include a range of conversion factors that define the value of a calorie:

To convert from	Multiply by	To obtain
calories, gm	4.1840	joules
calories, gm	4.18331	international joules
calories, gm, mean	4.19002	joules
calories, gm, mean	4.18933	international joules
calories, gm, 15°C	4.18580	joules
calories, gm, 15°C	4.18511	international joules
calories, gm, 20°C	4.18190	joules
calories, gm, 20°C	4.18121	international joules

And there are other standard conversions factors for calories, kg; calories, kg, mean; cal gm/°C; cal gm/gram; cal. gm/(gram × °C); cal gm/hr; cal gm (mean)/hr; cal gm/min; cal gm (mean)/min; cal kg/min; cal gm/sec; cal gm (mean)/sec; cal gm/(sec × cm²);   cal   gm/(sec × cm² × °C);   cal gm/cm²; and even

$$\frac{\text{cal gm–cm}}{(\text{hr} \times \text{cm}^2 \times \text{°C})}$$

If that is not complicated enough, there is now international agreement to abolish the Calorie altogether, and replace it with the joule. This value of the amount of work that is equivalent to a unit of heat was worked out by a brewer, James Joule, in 1843. The joule, now known by the abbreviation *J*, is defined as the amount of work done when a force of one newton produces a displacement of one meter in the direction of the force (or 10^7 ergs). A newton is a measure of the force which, applied to a free body having a mass of 1 kg, would give it an acceleration of 1 cm/sec/sec, and is equal to 100,000 dynes. A dyne is defined as the force that would give a free mass of 1 gm an acceleration of 1 cm/sec/sec, and an erg is the work done when a steady force of one dyne produces a displacement of one centimeter in the direction of the force.

All of which means that, whereas you used to know that 100 gm sweet corn provides 50 Calories, you ought in future to think of it as 50 × 4,184 joules. In round figures, multiply the Calories by 4000 to get the new answer in joules. And at the same time, you will now understand why I and many of my colleagues simply hate being asked 'What is a Calorie, anyway?'

To be more helpful than I have been in the first part of this answer, here is a list of some of the foodstuffs that contain so few Calories that you could eat them all day, and still not approach the level at which you would put on weight. Celery is one example, for you would have to eat 1000 sticks of celery to provide you with enough dietary intake in a day to keep you alive!

List of Non-Fattening Foodstuffs

Rhubarb	Cucumbers
Celery	Bell pepper
Cabbage	Radish
Endives	Strawberries
Lettuce	Cauliflower
Spinach	Tomato
Sprouts	Olives

List of only Slightly Fattening Foodstuffs

Asparagus	Apples
Broccoli	Melon
Beetroot	Orange
Carrots	Sweet Corn
Onions	Parsnips

Some interesting facts to remember are:
+ Butter is no more fattening than regular margarine
+ Potatoes are *less* fattening than chocolate, avocados, raisins, nuts, sponge cake, beefsteak, pork, or a frankfurter sausage. 100 gm of chicken meat rate 140 Calories, 100 gm potatoes a mere 90 Calories!
+ Rice is low in Calories, too. A portion of rice, at around 50 Calories, contains less energy than the average apple.

HOW IS FOOD DIGESTED? WHY DO WE NOT DIGEST OUR OWN STOMACHS?

Food has to be converted from fragments of chewed steak and moist remains of masticated bread and butter into absorbable chemical substances, and this is why food is digested. In the digestive system (digestive tract is the accepted term) the food is systematically attacked by a large number of enzymes. Some authorities speak of enzymes as though they were living organisms. 'Yeast is a well-known enzyme: it makes fermentation occur,' says one popularly-consulted work. The confusion arises because the Greek for 'yeast' is *zyme*, and since yeast produces fermentation through its own secretions, it was natural that the active agent would be named an 'enzyme' – meaning, literally, 'in yeast'. And (although yeast is a living microbe with a life of its own) the enzyme it produces is not in the least alive. Enzymes, since they were first discovered, have been found to exist in innumerable forms and they are nothing more mysterious than digestion-causing chemicals produced by living cells. They break down complex molecules into simple ones, and so an enzyme washing-powder is one that will digest organic stains that a mere detergent would leave behind.

What happens during digestion of food is that the materials pass through the digestive tract from mouth to anus, a range of enzymes attack the food, are mixed in with it, and the components are progressively and systematically broken down into a form that can be absorbed by the bloodstream.

The process begins with the saliva. Not only is food masticated and broken down into a pulp by chewing, but it is mixed with saliva which contains an enzyme known as ptyalin, or more accurately, amylase. Amylase breaks down starch in the food into sugars such as maltose, and it is easy to demonstrate this action. Chew a piece of bread for a few momements, and you will note how the sweet taste of the sugars gradually emerges as you chew.

The moistened and partly-digested food passes down to the stomach by muscular waves. It does not descend by gravity, and you may demonstrate that easily enough by standing on your head when you swallow the fragment of bread from the previous paragraph: it descends to the stomach – or ascends, perhaps, in the circumstances – even though it has to move against gravitational attraction.

The stomach is lined by some 35 million gastric glands. These pour out a welcoming bath for your food containing a mixture of enzymes, a good deal of mucus, and a strong solution of hydrochloric acid which is powerful enough to burn a hole in the carpet. It was in the 18th century that the French scientist René Réaumur (1683–1757) and the Italian priest-biologist Lazzaro Spallanzani (1729–1799) showed the powerful action of gastric juice on pieces of swallowed meat. Réaumur fed meat in little tubes to animals and birds, whilst Spallanzani swallowed morsels in perforated tubes on the end of a thread and pulled them out at intervals to see what was going on.

Both of them concluded that digestion was due to a process of chemical attack. But it was in 1819 that the first direct observations were made. A French-Canadian hunter, Alexis St Martin, received an accidental shot-gun wound to his stomach whilst hunting at Fort Mackinac on the shores of Lake Michigan. He was placed under the care of an American army surgeon, William Beaumont. Under the care of this young doctor, St Martin recovered, but there was a deep wound to the stomach itself which did not close. Through this, Beaumont was able to observe digestion going on and in this way he established much of our knowledge of the digestive processes in the stomach.

From the stomach, the food passes on through the small intestine, a tube like a fat hose-pipe about 21 feet long. It is divided into three zones, the first ten inches being called the duodenum, the next part (a little less than half of what is left) being the jejunum and the last the ileum, which is the narrowest portion of all. In the small intestine as a whole what basically happens is that digestion is completed, and the food residues are then absorbed by the wall of the small intestine. The most important source of digestive enzymes is the pancreas, a gland most of the public regard as a complete mystery. It produces a range of enzymes, and they enter the small intestine a few inches below its origin at the lower end of the stomach.

Note that although the food materials have been absorbed as they pass through the small intestine, none of the water that has been taken in with the diet has yet been extracted. There is a reason for this. The whole of the intestine is in a constant state of movement, squeezing, opening and closing, pulsating, writhing around to mix and blend the food with the digestive enzymes. The movement is altogether more violent and muscular than the coordinated and rather gentle 'waves of peristalsis' you learn about at school, and of course it has to be, if all that food is to be effectively handled. The more liquid the contents of the intestine are at this stage, the better; hence the retention of the water.

But in the final five feet of intestine, the lower bowel or large intestine, the water is absorbed until the faeces are reduced to a pasty mass. In the normal individual this is soft enough to be expelled with ease, but not so soft as to emerge ahead of its cue. Large numbers of bacteria (which may help to produce vitamins for us, and which almost certainly help to keep us healthy) grow whilst the food is passing through the lower reaches of the intestinal tract. One-third of faeces is actually composed of solid bacteria.

Now for some myths. Firstly, it is not necessary to evacuate the large intestine on a regular basis, every morning at 8.15, say. If this form of obsession suits you, then there is no harm in it, but the body is not evolved for acting in such an externally-mediated way and it is better to let matters take their course. Secondly, it is not obligatory for all those stages of digestion to take place in the complete sequence. If you read the textbooks you will find much detail about the vitally important action of all that hydrochloric acid and the role of the enzymes of the stomach, but there are people who suffer from *achylia gastrica* – a complete absence of gastric secretions – but they can enjoy perfect health, show no signs of impaired digestion, and never have any symptoms from the stomach at all. Thirdly, do not imagine that a 'furred tongue' and 'bad breath' have the least connection with the contents of your bowels. If you are in a sad and depressed state,

then you may develop such symptoms; if your body tone is flabby, your morale drooping, and your *esprit de corps* little more than a memory, then your functions may slow up in concert as a response. But nothing in the bowel *produces* bad breath or a white, coated tongue.

Finally, why does the body not digest itself? Surely it ought to, with all those enzymes acting one after the other and breaking down the food. There are two reasons. The first is the chemical makeup of the environment created inside your digestive tract. For instance, it has been found that the wall of the stomach produces a small amount of ammonia. This, being an alkali, protects the stomach lining from the effects of the acid (p. 79). In this and in other ways the lining of the digestive tract may be protected from harming itself through its own secretions.

In addition, the whole point of digestion is that the enzymes act one after the other, like a production line. If all the enzymes were present in the same area then doubtless the lining of the intestine would be attacked. But they are not; and each region can be considered to be specially adapted to resist its own characteristic digestive principle, even if it might not be able to resist the action of all the others.

We see the effects of this process in ulcers. Here the normal protective mechanisms break down, or there may be too much gastric juice produced, so that it can seep past the far reaches of the stomach and act on tissues that are not adapted to resist this particular combination of powerful digestive agents. When that happens, the lining does indeed begin to digest itself.

The intestine provides means of limiting the rate at which food is absorbed, and thereby helping lazy and overweight individuals to cut down on their food absorption without any need to cut down how much they actually eat. One of these relies on the use of a tapeworm, which is introduced in embryonic form into the intestine, and left there to grow. It competes for food with the host, and gradually the weight of the subject drops. When it is at the desired level, medicines are given to kill and expel the tapeworm. It is not unusual for teeth from the tapeworm head to remain lodged in the lining of the intestine, which can cause secondary problems, but the method has been used in America to pander to the needs of overweight and overindulgent individuals with a well-developed sense of the absurd.

An alternative approach is to remove surgically a large portion of the intestine, thereby hindering the absorption process. This mutilating operation has acquired some popularity in Japan. However, the desire of overweight people to offer themselves for such assaults does not offer much reassurance that they are ready to face up to the realities of life, and it could be argued that a doctor or dietician who counsels this road to slenderness is tending to sidestep the main issues. Whether that is good for the people who go in for this kind of treatment is hardly for me to say.

WHY DO WE EAT VITAMINS? ARE VITAMIN TABLETS USEFUL?

We have to eat vitamins because, if we did not, we should soon die. That is the bad news. On the other hand (and this is the good) there are more than enough vitamins in a modern diet to keep us robustly healthy and well-supplied.

Vitamins were discovered in the early years of the 20th century. Before then, it was imagined that a diet rich in carbohydrates, proteins and minerals would guarantee fitness unless some other disease agency supervened. But it had already been discovered that, in some strange way, citrus fruits (like limes) would help to protect sailors from scurvy – which is why Americans still remember the term 'limeys' for the British sailors of the 19th century. The notion that there might be some 'hidden ingredient' that earlier workers had missed was greatly strengthened in 1906 when Sir Gowland Hopkins of the University of Cambridge wrote the prophetic words:

'No animal can live upon a mixture of pure protein, fat, and carbohydrate; and even when the necessary inorganic material is carefully supplied, the animal still cannot flourish.'

The nature of these substances was unknown, and so they attracted the name 'accessory food factors'. Later they were thought to be all examples of the chemical class of compounds known as 'amines', and in 1912 the dietician Casimir Funk wrote:

'These deficient substances we will call "vitamines", and we will speak of a scurvy or a beri-beri vitamine, which means a substance preventing the special disease.'

It soon turned out that all the mystery molecules were not 'amines', and so the term was shortened slightly to 'vitamins' by 1926 and it is this term that we have inherited. The subject of what exactly a vitamin is, and how many you could claim existed, is far from final resolution. But some idea of the extent of the vitamin league table is shown by the following list. Of greatest interest is the way that a given entity – like vitamin B, for instance – has in the fulness of time turned out to be a complex range of substances, each with a different structure and a distinctive role in the metabolism of the body.

An Outline of the Vitamins

Vitamin A
    antixerophthalmic vitamin
Vitamin B group
    vitamin B_1 = thiamine
    vitamin B_2 = riboflavin
    nicotinic acid = niacin
    vitamin B_6 = pyridoxin
    pantothenic acid
    folic acid
    vitamin B_{12} = anti-pernicious anaemia factor
    biotin
    inositol
    para-aminobenzoic acid
    choline
Vitamin C, ascorbic acid = antiscorbutic vitamin
Vitamin D, 7-dehydrocholesterol and
    calciferol = antirachitic vitamin
Vitamin E, α-, β-, and γ-tocopherol = antisterility
    vitamin
Vitamin K = antihaemorrhagic vitamin

Note: Vitamin G is an earlier name for what we now know as riboflavin, vitamin B_2.

The effects they exert are many and varied, and they cover a large spectrum of vaguely-defined 'vital functions'. Some of the vitamins are so abundant that we know nothing of the effects of a lack of them in man. Among those for which deficiency has never been known are pyridoxin (B_6), pantothenic acid and para-aminobenzoic acid. The effects of a lack of biotin are known only by experimentally depriving volunteers under laboratory conditions.

Vitamin A helps maintain the tissues, particularly the skin and other epithelial tissues, and is essential in the visual cycle where it helps regenerate the pigment known as 'visual purple'. Without it, skin diseases and night blindness develop. The main source of Vitamin A is carrots, which is why you are told to eat them so that you can see in the dark. Only if you are already a victim of severe deficiency will that promise come true.

Vitamins of the B group are plentiful in a number of foods, most notably yeast and liver. All of them are concerned with enzymes in some way or another, and an inability to absorb B_{12} results in the development of pernicious anaemia, a debilitating and progressive disease that was once untreatable, and in the early decades of this century could be ameliorated only by feeding the patient large amounts of raw minced liver. Not a pleasant prospect.

Vitamin B_{17} has recently entered the news, and is apparently nothing of the sort. That is, the substance is a toxic extract of almonds, and is certainly *not* a vitamin. Some people claim it will produce miraculous cures of dreadful diseases, but all the evidence I have seen so far suggests it is little more than a profitable side-line for the retailers. In any event, B_{17} is not a vitamin and nobody can explain to me how it gained its preposterous name.

Vitamin C, ascorbic acid, is involved with the maintenance of the substances that hold many of the body's cells together. Without vitamin C, the blood vessels, for example, start to come apart and leak. The skin becomes mottled with broken blood-vessels, the heart muscle is damaged, there is bleeding into joints and the gums may bleed. It was a wretched condition until the simple answer

– adequate supplies of fresh fruit and vegetables – was recognised. You still find cases in babies on artificial feeds, if they are not given fruit juice, and in trappers or lumberjacks who may live out of contact with fresh fruit for months at a time. They ought to take vitamin C supplements as a matter of course.

Vitamin D deficiency results in rickets, since the compound is involved in the metabolism of calcium, and rickets is typified by weak bones and teeth. The two important varieties of vitamin D are D_2 and D_3 (there is no vitamin D_1) and an important natural source is the human skin! Sunlight on our bodies converts naturally-occurring substances in the skin into vitamin D; indeed it was in 1880 that the value of sunlight in the treatment of rickets was first noted. (However, it took almost half a century before the claim was generally accepted.) Vitamin D is found in fish-liver oils, but the best source is a pleasant, sunny day.

Vitamin E, the antisterility vitamin, is found in wheat germ and water-cress, and also in lettuce leaves. A total lack of this vitamin results in

... what lettuces can do for rabbits they can do for you ...

sterility. But that does not mean that a large dose will produce a raised libido! Large amounts of this vitamin are consumed each year by waning suitors who hope for a miraculous rejuvenation at their nether end, but unless they are in the throes of a deficiency – which is vanishingly improbable in the western world – their hopes seem to be in vain. Vitamin K is involved with blood clotting, and a lack will lead to haemorrhagic disease. It is rarely seen in man, though it has been found in farm birds reared on a restricted diet. 'K' is found in many green crops; alfalfa is the example everyone quotes.

The vitamin you need to worry about is ascorbic acid, vitamin C. This is because you need larger amounts of this than any other, and because the body cannot store it as is the case for most vitamins. So regular supplies are necessary. But you do not have to eat endless supplies of citrus fruits to obtain your vitamin C. In most normally balanced diets there are large amounts (by this I mean an amount that is more than sufficient to keep scurvy at bay), and there are other sources that you rarely hear about. There is at least as much vitamin C as there is in an orange (if not more) in broccoli, peppers, and parsley.

Does vitamin C in larger amounts protect against the common cold? The evidence does little to support this view, but in spite of that I tend to take a few grams in the form of effervescent fruit-flavoured tablets whenever I imagine that a cold is due, and seem to suffer colds very rarely. Do not imagine that everything about science has to be scientific, for most things aren't; the fact that I take vitamin C – and almost believe in it – in the face of the evidence is a parallel to the story of the physicist who was seen to nail a horse-shoe on his laboratory door.

'What is that there for?' enquired a visitor.

'It is a token of good luck,' said the physicist.

'But I did not think you were superstititious, and believed in old tales like that!' came the incredulous retort.

'I'm not superstitious!' protested our physicist friend, and then added, after a thoughtful pause: 'But I am assured it works, even if you *don't* believe it.'

The vexed question of what vitamin pills people should take would require a set of full-length books to analyse in detail. I have said that diets these days are unlikely to be deficient, and that is true; occasional examples of scurvy in the backwoods and the resurgence of cases of rickets in the elderly or nursing mothers in under-developed countries do not affect the general rule. Sufferers from deficiencies need urgent and prompt treatment, but most of you who read these words will not come into that category. Vitamin pills are probably a complete waste of money and effort, but taken in sensible doses they are unlikely to do any harm, and – who knows? – just may do you a little good. Even if (like the physicist with the horse-shoe) you don't really believe in them.

SHOULD MANKIND EAT A VEGETARIAN DIET?

No. Many vital dietary constituents, like vitamins, are normally found in animal tissues, and though you can obtain them from other sources there is no 'naturalness' in doing so. The teeth of our species reveals that we are meant to eat both animal and plant matter. Carnivores like lions and dogs have a closed root to the tooth; and the teeth themselves are jagged in order to tear the flesh of the victim apart. Herbivores, like sheep and cows, have flatter teeth with an open root, plentifully supplied with blood-vessels to nourish the tooth as it grows. Humans have typically intermediate teeth, with narrow – but not closed – roots to the teeth, and a halfway-house between the herbivore and the carnivore shape for the crown. Similarly, the intestinal tract has an intermediate structure (in hervibores, the appendix – which so often causes us problems – is a large extension of the tract for digesting vegetation, whilst in pure meat-eating species it is absent altogether). Man must be regarded then as an omnivore, one who eats both plant and animal material.

If there is little justification for the idea that vegetarianism is somehow 'natural' – for it is not – that does nothing to say that being a vegetarian is pointless. It is true that the means by which we farm and slaughter animals causes them discernible distress, and you could convincingly argue that we ought to know better. The inhumanity in the intensive rearing of animals is an entirely valid standpoint from which to argue, and if people choose to adopt the vegetarian diet as a response to that, and as a gesture towards a more gentle and considerate world, then no one should imagine that is foolhardy, for it could be a saving grace.

ARE 'NATURAL FOODS' BEST?

They are if you are living a 'natural life', which would entail hunting with sticks and stones and living behind a bush. But most of what we regard as 'natural' in the modern era is nothing of the sort. The wholemeal bread we treasure as the mainstay of a 'natural' diet represents the height of sophisticated food technology. The grain from which the flour is milled is a non-natural crop, which has been selectively bred over thousands of years from wild grasses. The grinding is unnatural, for nothing like it can happen without mankind's technology, whether it takes the form of a modern mill or a hand-carved grindstone driven by hand. The ingredients are then fermented by yeast, which may be a time-honoured process, but is none the less a version of what it is now fashionable to call biotechnology. During this process the bubbles of the carbon dioxide gas that are given off as the yeast metabolises make the dough swell up into a foamy mass. The loaf is then cooked, at a carefully maintained temperature unlike anything you would find in nature (except perhaps near a volcano) until the centre is cooked and the outer layers are caramelised and brown, but not burned.

This is a well-tried – but wholly artificial – scientific process, and the product we relish as the symbol of conscientious eating has nothing to do with nature. What man managed to do was to harness the life processes of the yeast microbe to turn an unpalatable material, powdered seeds, into a light-textured and attractive foodstuff, bread. The practical advantage of using yeast is not only that it makes the dough rise by the production of carbon dioxide bubbles. The enzymes that the yeast produces undertake some predigestion of the carbohydrates present in the flour. And when the bread is cooked, the action of the heat converts the somewhat soft and slimy growth of yeast into a firm and almost rubbery consistency that gives the loaf its resilient texture. You can easily witness this effect of heat on proteins in cells by frying a morsel of raw liver. In the fresh state the liver is very soft and has a slippery, slimy texture. But as the heat of the frying pan or skillet acts upon the liver, the protein in each cell becomes denatured: it changes from being soft, wet material into a much harder one. Another example of the denaturing of protein is the cooking of egg white, which is soft and slimy in the natural state, but resilient and almost rubbery when denatured by heat. The comparison between raw and cooked egg-white exemplifies the difference between the wet, fresh yeast in risen dough, and the cooked yeast in a baked loaf.

When you have your loaf of wholemeal bread, what do you spread upon it? Why, margarine, of course. This must be one of the biggest confidence tricks of the whole dietary arena. Everyone is told that margarine is less fattening, and somehow more 'natural', than butter. In fact margarine contains just as much fatty material, it is the equivalent of just as many Calories as butter, and it is not even exclusively plant-based. Animal fats are used in magarine manufacture too. So, unless you purchase a product that has a declared reduction in energy content – which some do by whisking water into the product – or which is guaranteed to be free of animal constituents, you are not buying what you may think you are.

It is at this point that the butter manufacturers like to proclaim the essential naturalness of their product. But this will not do, either. Nothing could be less natural for a human than drinking the secretions of the under-belly of a cow, even if the secretion (which we call 'milk') has been churned to make butter! The advertisements for butter that you can see, which emphasise that butter is 100% pure, with just a little salt added for the flavour, are perpetrating a fallacy. Butter is often coloured with dyes to make it a bright,

'. . . less fattening . . .'

marketable yellow, and unadulterated butter is something you have to search for, not something you find piled high in the world's shops.

What might you eat with your bread and margarine? Cottage cheese perhaps – that alien milk, once again, processed by microbes – or possibly some pickles – vegetables preserved in vinegar. You might eat some yoghourt, which is another artificially processed food that cannot exist in nature. After this, a can of baked beans or a packet of potato crisps looks like naturalness itself.

The campaign for 'natural' foodstuffs is a hollow sham. Neither should you be convinced that traditional foods are necessarily better. The most widespread of all the staple tropical crops is cassava, *Manihot utilissima*, known in temperate countries as the source of tapioca. Yet cassava can poison consumers with its natural content of cyanide.

Cassava (sometimes known as manioc) was originally a native of America. It has woody stems about six feet high from which spring umbrella-like leaves, and a swollen underground root system which is the part that is harvested and eaten. Cassava is grown throughout the tropics where there is a rainy season, and it has several important advantages for the farmer. One of these is that its woody stems make it resistant to locust attacks, and cassava plants can recover from a locust swarm that would destroy a lesser crop.

Its other characteristic feature is that the tuberous roots can be left in the ground where they have grown for two or three years without further attention, and without being damaged by pests. Its main disadvantage is that its food reserves are almost exclusively in the form of starch, so that although it is rich in energy it contains too little protein to make a balanced staple food. However, cassava in the ground is a tropical standby as valuable as a well-stocked granary in temperate latitudes, and for this reason alone it deserves to be better known.

There should be no surprise about the fact that the cassava root remains underground for prolonged periods without being attacked by parasites. The plant is a potent source of cyanide, and would kill almost anything that tried to eat it. The cassava plants which pose the greatest risk are the bitter-tasting varieties, and there are sweet cassava strains that are less hazardous. However this is not an unvarying property, as the types tend to change from bitter to sweet and back again in response to the climate, and all forms of cassava are potentially deadly.

Fortunately, the cyanide is destroyed by cooking, and so cassava is always cooked before being consumed. The most popular way of doing this is to peel the roots, boil them well and mash them like potatoes. The roots can also be grated to produce a meal (such as the 'garri' of central Africa, and the 'farinha' of Brazil) which is then pressed into round cakes and cooked as a traveller's snack. The juice can be fermented to make alcoholic drinks, or boiled down to make an extract (known as 'cassareep' in the West Indies) which is used in sauces. Tapioca is made in flakes or rounded grains, known as pearls, by drying the starchy product of the ground-up roots. Even the leaves can be consumed by being gathered as a crop and cooked like cabbage.

So cassava can be seen as a vital part of the economy of the underdeveloped nations. It is a staple item of food for millions who would die without it. Yet unless it was cooked (and peeled, for the skin of the root tubers contains the highest concentration of the deadly poison) it would kill. Let us carry the example of the cyanide-rich cassava in mind whenever we try to convince ourselves that 'naturalness' is good for people.

The only truly natural food any of us take is mother's breast milk, apart from the direct input through the placenta we obtained before birth. Even these are not always good enough. Babies can be born so deprived of the blood clotting vitamin K that their lives are in danger. If they survive without medical intervention then within a few hours they acquire a population of bacteria in their intestines which manufactures a supply of vitamin K, and this saves the baby's life. Mother's milk is not all it is cracked up to be either, for it is deficient in vitamin D. In sunny climates, babies are able to produce the deficit in the skin through contact with sunlight (see p. 27) but when a baby

is dressed in clothes to keep it warm, or even wrapped in a blanket, then the amount of vitamin D it obtains through its mother's milk is generally not enough. It is from this fact that the unsavoury habit of feeding youngsters spoonfuls of cod-liver oil was developed.

To sum up, most of the foods we think of as essentially 'natural' are nothing of the sort. Some foods that really are natural are either deficient, or else deadly. One answer to this apparent conundrum is that being civilised is a highly unnatural state, and we need carefully considered – but artificial – means to survive in it. There is no virtue in trying to avoid processed foods, for nothing could be more processed than bread and margarine, cheese or yoghourt. Neither is there any point in trying to avoid 'chemicals'. All food is made of chemicals, and so too is the person who eats it.

The guiding principle in food preparation should be to avoid putting in ingredients that are of questionable value or of unknown effects. For example, I have by me as I write a jar of imported green beans which contains nothing but beans, water and a trace of salt. Alongside them is a container in which the beans have been coloured and enhanced in flavour. Of the two, the first example is preferable to the second. If a food contains additives to keep it healthy, or to boost its nutritional value, then it may be an excellent diet. Sometimes, however, foods look as though they are acting as an 'elephants' graveyard' for a range of unwanted chemicals produced by the industrial companies with nowhere else for them to go. I hope that is not an uncharitable attitude, but it sometimes seems as though wholly superfluous additions are being made to food better left alone. For example, we spend vast amounts of money stripping the bran layer from wheat grains and grinding white flour from what remains. This leaves the bread textureless, and the latest trend is to restore some of the original fibre component to the flour by mixing in a percentage of ground-up pea-pod shells. At times like this, one wonders why anyone bothered to go to such lengths removing the fibrous component in the first place, if it has to be substituted like this. Much of the bran ends up as breakfast cereal, in any case!

I think food manufacturers should adopt a new *principle of minimum interference*. If the food is good enough as it is, why spend time, money and effort interfering?

IS IT HEALTHY TO CHEW EACH MOUTHFUL 32 TIMES?

Certainly not. This is an obsessive and silly Victorian saying. Try doing it with a mouthful of custard!

HOW DOES A CAMERA WORK?

It is surprising how many people take pictures without knowing anything about how the process works. The lens, at the front of the camera, is easy to demonstrate. Any magnifying glass, held a few inches in front of a piece of paper, will focus an image of the view it faces. If you try this in a living room, a clear image of the window, or of the room lights if it is night time, can easily be obtained on a piece of paper by holding the manifying glass in contact with the paper and then moving it away until the image becomes clearly focused.

Of course, you have to control the amount of light that passes through the camera lens, and there are two ways of doing this. One of them involves altering the diameter of the hole through which light can pass, known as the aperture or the stop; and the other alters the length of time that the lens shutter is opened. The easiest lens in which to observe the effect of aperture change is your own eye. If you look at a bright window or an electric lamp then a companion will see that the pupil in the centre of your eye (the aperture through which light enters) is very small. If you now shade your eyes from the direct light, your companion (or you, if you try this with a mirror) will see the aperture widen as the light level falls.

The shutter mechanism needs no demonstration. A camera shutter opens and allows light to enter the camera for a set length of time, as though you flicked your eyes open and shut again in a fraction of a second. The important aspect of taking photographs is to ensure that the right amount of light is allowed to reach the film. In professional cameras, both the aperture and the shutter speed are varied to obtain the desired result, whilst in simpler snapshot cameras one of the functions (usually the shutter speed) is kept constant, and there are several settings for the aperture to take account of different lighting levels. The simplest cameras of all have both fixed, with a shutter speed that is short enough to help overcome camera shake and an aperture that is sufficiently small to give good picture quality over a considerable distance.

It is important to realise what effect these two factors have on the quality of the picture. A short shutter speed freezes movement, whilst a longer one will allow moving objects to blur. An open aperture allows in more light, but limits the depth of the field that is seen in focus. This means that the object on which you have focussed will be sharp, but objects nearer to the camera or further away will be out of focus. If the lens aperture is closed, or 'stopped down', then the amount of light is reduced but the depth of field of the image is considerably increased. If the aperture is opened in order to obtain the shallow depth of focus effect for artistic purposes, then the shutter speed will have to be increased to compensate.

By using the appropriate settings, therefore, you can obtain an identically exposed pair of pictures by using a wide aperture with a short exposure, or a small aperture with a long exposure. Which you would chose depends on the subject: if it was vital to freeze the movement of an athlete, for instance, then the widest aperture would be selected and the shortest shutter speed. You would have to take care that the focus was accurately set, for even a slight discrepancy could throw the image out of focus. Conversely, if you planned a shot of a garden with distant and close features in view, you might prefer to select a shut-down aperture and compensate with a long exposure time. Here you would have to be careful not to use too long a setting, or you might obtain an image blurred by camera shake. Mounting the camera on the top of a tripod is the answer to this problem.

Many smaller cameras have no focussing control at all. This is usually the case for the miniature cameras that take 110 film. They are made with the lens pre-set for focus, usually at around 15 feet; and the lens aperture is set in a midway position so that the depth of focus will give reasonable pictures over a wide range of object distances. A camera like this needs to be used sensibly and not brought too close to the subject, but for normal purposes they can take excellent pictures.

WHAT DOES THE 'f' NUMBER MEAN?

The aperture of a camera lens is defined by its f number. A lens that is stopped right down, as every photographer knows, might be at $f/11$ or $f/16$, whilst a lens with a wide aperture would be at $f/2.8$ or $f/1.8$. The lower the number, the more wide open is the aperture. The maximum aperture is limited by the manufacturer's design. Most

Sometimes a second setting for a duller occasion

modern lenses open to $f/2.8$ but faster lenses, as they are called, open to $f/1.8$, $f/1.4$ and even $f/1.2$, with occasional professional lenses, made for specialist purposes, opening even more than this. The pre-set miniature snapshot camera with a fixed shutter speed will have a lens rated at around $f/8$, sometimes with a second setting for duller subjects.

And what does the 'f' itself signify? It is the accepted abbreviation for the focal length of the lens, so that a lens that focuses an image at infinity onto a plane 50 mm away is said to have a focal length of 50 mm. This would be written as $f = 50$ mm. So if you set that lens, say, to $f/2$ you would find that the diameter of the aperture is exactly 25 mm. The abbreviation is actually a fraction: $\frac{f}{2}$ or the focal length divided by 2. In the earliest days of photography the aperture was often spoken of in terms of its diameter in inches or fractions of an inch, and the f convention was introduced because it was realised that the actual size of the aperture was not as relevant as the size of the aperture in relation to the focal length of the lens. If you had a half-inch diameter aperture fitted to two different lenses of disparate focal lengths they would both behave differently. But an aperture of $f/2$, i.e. half the focal length, admits the same amount of light no matter to which lens you apply it.

The stop numbers around the lens of a camera seem to be a strange assortment: 16, 11, 8, 5.6, 4, 2.8 are the usual figures. The secret of this strange sequence is that each figure, when divided into the focal length of the lens, provides an aperture area that is progressively halved each time. Thus, for a 50 mm lens, a stop of $f/11$ is an aperture area of approximately 50 sq mm, a stop of $f/8$ doubles the area to approximately 100 sq mm, and $f/5.6$ to approximately 200 sq mm. The approximate doubling up continues through the other settings, $f/4$, $f/2.8$, $f/1.8$ to $f1.4$, which has an effective area of 3000 sq mm. The next fastest stop found in currently produced lenses is $f/1.2$, but this is still slower than the human eye's sensitivity. To match that, you would actually have to go to $f/1$, but no generally available lens is made to this specification.

You can now see why it is that very fast lenses

33

are not readily obtainable. To make an $f/2$ lens of focal length of 50 mm would mean that the maximum lens aperture (which is roughly the same as the diameter of the lens glass itself) would be 25 mm. That is an easy task to undertake, from the manufacturer's point of view. But a single lens built on the same principles with a focal length of 1000 mm, a telephoto lens, would need to be 500 mm in diameter to rate at $f/2$. It would weigh, figuratively speaking, a ton. Long focal length lenses like this either have a much reduced aperture, say $f/8$, or they are made with concave mirrors which can be made easier to transport.

In practice, the longer the focal length of a lens the more it magnifies the subject, and therefore the more it magnifies any movement in the system. A wide-angle lens with a focal length of 24 mm allows you to use very long exposure times with a hand-held camera, whilst for anything more than a 135 mm telephoto lens the camera needs to be mounted on a tripod. Is there a rule of thumb to know what to do? There is one I use, and it works very well in practice even though it is an arbitrary relationship. Use the shutter speed nearest to the focal length of the lens if you are hand-holding the camera. Thus do not use anything longer than one-fiftieth second with a 50 mm lens, or one-five-hundredth with a 500 mm lens.

Incidentally, the newcomer to photography will notice that the shutter speeds are multiples of 2: typically 1 sec, $\frac{1}{2}$ sec, 1/4, 1/8, 1/15, 1/30, 1/60. 1/125, 1/500, 1/1000. There are slight discrepancies to make the mathematics easier – 1/30 is not half 1/16, any more than 1/60 is half 1/125th – but these are of no practical consequence. The principle is that if you set a camera on a shutter and aperture that will give you a perfect exposure (say, 1/60th at $f/11$), then if you change the shutter speed by two settings one way, you will still have an identical exposure if you alter the aperture by two settings to compensate (i.e. 1/500 at $f/5.6$). What you have done by the change is to quadruple the amount of light passing through the lens, whilst quartering the amount of time the shutter is open. The result in terms of exposure should be the same.

WHAT ABOUT AUTOMATIC CAMERAS?

Automatic cameras have a light sensing device that controls the exposure electronically. Some systems take account of the contrast of the scene, whilst others measure only brightness and have to be manually over-ridden to take account of large light or dark areas that might otherwise 'fool' the sensors.

Most automatic cameras set the shutter speed automatically. The simplest kind have a set aperture, perhaps of $f/8$; they measure the light reaching the camera and then the shutter is closed a set interval after you have manually opened it. A light goes on in the viewfinder if the exposure is going to be too long for the camera to be hand-held (longer than 1/30 second, typically). A flash unit is then recommended, though I prefer to set the camera down on a firm surface like the top of a cupboard or a nearby shelf, and hold it down whilst pressing the shutter. In this way pictures can be taken by indoor lighting, and if your subjects will keep still long enough, portraits of people in a dim evening-lit room are feasible.

More sophisticated cameras display the selected shutter speed before the picture is exposed. Then, if the speed is too slow, a wider aperture can be chosen. One refinement measures the light reflected from the film plane, which is ideal for use in technical applications or when light levels are unpredictable. Unfortunately, this system only measures overall brightness of the image, so compensation has to be made for unusual scenes – such as a dark-dressed person against snow, which needs overexposure by 2 stops, or a scene lit by a single bright lamp, which will need overexposure by $1-1\frac{1}{2}$ stops.

This system, in which the photographer selects the preferred aperture and the camera selects the shutter speed, is known as the aperture-priority system. Other cameras with an automatic facility allow you to set the shutter speed, and the camera then selects the correct aperture, a system known as shutter-priority. There are cameras with both facilities, and the user chooses which is best for the circumstances; and some have a modular system which allows you to select a desired combination, and the camera will then modify either setting or both, depending on light levels detected. It is probably true to say that the automatic shutter – i.e. aperture-priority – is most useful for general purposes, though the shutter-priority system can be valuable for people who wish to photograph fast-moving subjects in sport, etc.

Many users find that the more complex cameras are so over-equipped with gadgetry that they become inconvenient in practice. Professional photographers like to keep control of their machinery, and with too much automation the machinery seems almost to control them.

HOW DOES THE PICTURE GET ONTO THE FILM?

The focussing of the picture and the setting of the exposure values gives us an image projected on the film carrier at the rear of the camera. As the focussing experiment on p. 32 showed, the image is inverted. There is a mirror prism inside a single-lens reflex camera that rectifies this and restores the image to its right-way-up appearance.

Typical black-and-white film consists of a plastic base (usually made of triacetate) on which is spread a layer of silver bromide crystals in gelatine. The silver bromide emulsion is prepared in darkness, and the crystals can be activated by contact with light. They do not immediately alter their appearance to the familiar black colour where light has struck, for the activated crystals need to be developed before the image will appear.

Developing agents (usually metol, hydro-quinone and phenidone) convert the activated crystals from the near-white colour of silver bromide to a black deposit of silver. In this way a bright object is recorded on the film as a black image, and a very dark object – which produces no alteration in the silver bromide crystals – is represented by unaltered, near-white emulsion. We have in this way obtained a negative of the original scene.

The unaltered emulsion will be affected by

light, and so the non-activated silver bromide crystals are dissolved out of the emulsion layer by a solution of sodium thiosulphate or an ammonium salt, known as 'fixer'. The fixed image is then used to prepare photographic prints.

The normal sequence by which a film is processed thus consists of (a) loading the film in the dark into a developing tank, (b) adding developer for a carefully calculated time (say five minutes, (c) washing under a running supply of filtered water, (d) fixing for a time that will vary with the kind of fixer used, (e) washing to remove traces of fixer that might lead to subsequent image degradation, (f) drying in a dust-free atmosphere. In some processes, an acidic stop-bath is used before stage (c) to arrest development.

Larger bromide crystals are more sensitive to light than small ones, and so fast films are made with these larger silver bromide particles. Slower films, made with smaller crystals, produce a better definition. In practice photographers know that the more sensitive the film, the coarser will be the grain of the image, and so for high-quality enlargements a fine-grain film is chosen.

Film made exactly as described would only be sensitive to the blue end of the spectrum. For this reason, other salts are often incorporated (silver iodide, for instance) and the emulsion may be treated with small additions of mercury, gold and other heavy metal ions, and also by sulphide ions. By the use of carefully chosen compounds, film can be produced that is equally sensitive to a wide range of colours.

Film speeds are described according to three systems, in each of which the larger the number, the faster the film. The most familiar for English-speaking users is the American ASA system. Here roughly doubling the ASA number indicates an approximate doubling of film speed (so that film of ASA 400 is twice as sensitive as ASA 200 film). The Russians employ the GOST system, in which the values are somewhat different but basically the same mathematical basis applies. In this case, the films described in the example above would be GOST 360 and 180. To all practical intents, the GOST and ASA systems can be used as virtual equivalents.

The German DIN standard does not increase

by doubling. In this case a doubled film speed is signified by an addition in value of 3. The examples we have considered above would be DIN 27 and 24. A table of the equivalent values is given below.

Film Speed Systems

ASA (American)	GOST (Soviet)	DIN (German)
16	14	13
32	27	16
64	55	19
125	110	22
200	180	24
400	360	27
800	720	30
1250	1125	32
2000	1800	34

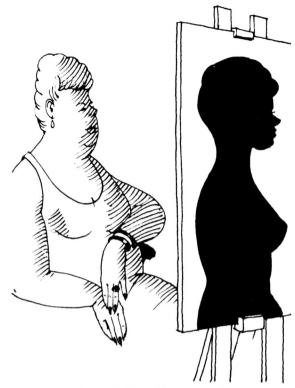

'. . .a fashionable means of recording the most faithful portrait . . .'

WHO INVENTED PHOTOGRAPHY?

The earliest claim for the discovery of photographic, light-sensitive materials was made by a Hong Kong scholar, Hun Tsi-kwan, who said he had unearthed plates made 2000 years ago which were coated with materials that would have reacted to light. Lenses, as burning glasses, have been known since 400 B.C., when they were recorded by the Greek playwright Aristophanes. The first camera (meaning a closed chamber, for it was not a device for taking actual photographs) was described by the Greek philosopher Aristotle in his book *Problemata* around 350 B.C. His camera did not have a glass lens, but focussed light from a scene through a small pin-hole onto a screen. The first time that images were captured from a projected scene like this was in the later years of the 18th century, when silhouettes – cut from the projected outline of the sitter – were a fashionable means of recording (in the words of the Swiss philosopher Jean Gaspard Lavater, 1741–1801) 'the most faithful portrait a man can have . . . a direct transcript from nature'.

In 1802 Sir Humphrey Davy wrote a paper describing some experiments in producing silhouettes and nature-prints by direct chemical means, the first true photographs as such. His paper was entitled *An Account of a Method of Copying Paintings upon Glass and of Making Profiles by the Agency of Light upon Nitrate of Silver, with Observations by H. Davy.* It was published in the *Journal of the Royal Institution* and discussed a series of experiments that had been carried out by Davy's friend Thomas Wedgwood (son of the founder of the great pottery company).

So photography and the camera had a long history before we even introduce the names of the people who are traditionally associated with the subject. The first photographic image of a view, focussed by a lens in a camera, was described by its inventor, the French army officer Nicéphore Niepce in a letter he wrote to his brother on 3 September 1824. He tried coating plates of metal with bitumen, and exposed them in a camera obscura for eight hours. The bitumen was hardened where sunlight struck it, and the remaining areas could be washed off and the bare metal so revealed etched in acid. In the museum of Châlons-sur-Saône (which was Niepce's native town) two of these plates were preserved with the caption: 'Dessin héliographique, inventé par N. Niepce, 1825.'

In January 1826, Niepce learned that another French experimenter, Louis Jacques Mandé Daguerre, had similar aims. They became collaborators in 1829, but three years later Niepce died before any of their work was published. Daguerre showed how mercury vapour could be used to 'develop' an image on a flat plate which could then be fixed in a solution of salt in water. Daguerre revealed his process publicly in 1839, and in a matter of weeks everyone wanted to be photographed by the new technique.

Meanwhile, an experimenter of independent means from Lacock Abbey in Wiltshire, William Henry Fox Talbot, was trying experiments of his own. In 1835 he succeeded in taking the first negative. The picture, the size of a postage stamp and pale lilac in colour, still exists. Alongside it are written the following words:

'Latticed Window (with the camera obscura) August 1835. When first made, the squares of glass about 200 in number could be counted, with the aid of a lens.'

I have examined the paper myself, and the little panes of glass are still clearly visible. There are well over 200 of them.

When Daguerre announced his discovery, on 5 January 1839, Fox Talbot was shocked. He wrote to France, claiming similar recognition for his experiments, but he was tersely informed that Daguerre's results 'excite the admiration of all our artists by their perfection and delicacy'. The truth is that Daguerre's method was in essence a one-off process. He used a plate of copper, coated with silver and bearing a film of copper iodide. This was exposed in the camera, and the resulting image developed by mercury vapour before being fixed in the salt bath. Fox Talbot was making paper negatives, which could be used to print positives. In that sense his process was the direct antecendent of modern photography. But Daguerre deserves his fame, for he was actively working towards a commercial proposition designed to bring the photographic image within reach of the public. Fox Talbot, more a man of leisure, had many other projects on his mind and worked mainly for personal interest.

In the years that followed, the daguerrotype swept across Europe and beyond, whilst Fox Talbot perfected a similar portrait process of his own, the calotype (a name derived from the Greek $\kappa\alpha\lambda\sigma\zeta$ = beautiful). During the 1840s this process recorded countless views and personalities (from the building of Trafalgar Square in London, to a study of the ageing Thomas Moore, who died in 1852). By the mid-1800s photography was commonplace, and the great Victorian photograph album – which has become one of the most valuable social records – had been established.

WHAT IS A MICROCHIP?

The microchip is a very small transistor circuit, with many different components of minute size. Transistors were made by sandwiching together specially prepared components which contain extra electrons with compounds where there are, as it were, vacancies for electrons. As electrons have a negative electric charge, the component with extra electrons is known as an n-type material. The component with vacancies – electron holes is how they are thought of in the trade, though there is no real 'hole' of course – are known as p-type (for 'positive'). If the two are sandwiched together you have a 'pn junction'.

The advantage of a pn junction is that if an electric current flows one way through it, from the n towards the p, the electrons will be carried into the p material where they then fit into the vacancies – the electron holes. If the current is connected up the other way round, however, electrons from the n region and electron holes in the p domain, as it is called, are attracted away from each other. In this instance there are no electrons moving into the holes, nor holes to accept them, so there is no flow of current through the entire transistor.

That may seem a complex way of switching current so that it can only flow one way. But this kind of process is known in electronics as rectification, and one-way flow of this kind could only be obtained before transistors by using glass vacuum valves which were costly, delicate and complicated to handle. The dawn of the transistor meant that tiny switchable units could be assembled into calculators, so that a computer (which is nothing but a huge array of binary switches, much like this) which used to weigh many tons, fill several large rooms and consume as much power as a passenger train, could now fit into something the size of a cupboard. By making more complex junctions, npn and pnp transistors were introduced, and although simple enough on their own, once they were wired into large arrays their capacity to undertake prodigious mathematical feats was greatly increased. The principal ingredient of a modern transistor is silicon, which is 'doped' with traces of elements such as phosphorus and arsenic to make n-type components or with an element such as boron, which lacks electrons, to produce the converse effect.

The step from the transistor to microelectronics and the birth of the microchip has been made because it was realised that the same effects were observed, no matter whether the unit itself was large or small. Instead of sandwiching together large amounts of n and p materials, it was argued that a whole circuit could be built up by adding layers to a single unit – a chip – so that you could have dozens of components in the space originally occupied by just one.

The company that began this race towards miniaturisation was Fairchild Inc., a major electronics manufacturer, and their Director of Research, Gordon Moore, rather boldly advanced the idea that from 1959 onwards, the number of components that could be fitted onto a single chip would double every year. That would mean two by 1960, four by 1961, eight by 1962 . . . until we had 2,048 by 1970 and over half a million by the end of the 1970s. He was not far wrong. Even though his theory was proposed in a bold and improbable way, by 1979 it was already possible to make a chip with 250,000 components which is gratifyingly (you might even say 'surprisingly') near the predicted level. I do not know his thoughts on the matter, but I guess that Dr Moore would be surprised if anyone had told him in the

early days how popular the microchips would become. In that year, 1979, the total number of chips sold had reached the total of thirty million.

The first tiny computer to enter the market was a microprocessor made by the Intel Corporation of the USA which was announced in 1971. It had the equivalent of 2,000 transistors mounted on a chip the size of a finger-nail. Yet this diminutive device had the calculating capacity of the ENIAC computer, which weighed 30 tonnes and took up as much floor space as a modern household. Whereas the engineers who worked on ENIAC used to walk about it, checking valves and wiring, the modern electronics engineer is working with microchips on which the components are nearer the size of bacteria. The fall in size has led to a drop in cost. Since the late 1960s the cost of a given memory unit has fallen by 30% per year.

Of course, though they have given rise to such useful luxuries as calculators and the digital watch, microelectronics is not going to solve all our problems. Many systems where computers have been introduced are far less efficient than they were. The problem is that planners are not always as worldly as they might be, and they tend to find themselves so consumed by what they are selling – the system – that they lose contact with the world whose problems we are trying to solve. One datum they love to quote is that computers can handle vast amounts of information: if present rates continue, they say, then data will be generated by 1990 at such a rate that if they were printed on paper, the pile of pages would be as tall as Mount Everest in just thirty minutes.

Now, that may be a wonderful thing to an individual who spends his life celebrating the majesty of mega-mathematics. What worries me is that the capacity to handle all this information will encourage people actually to do it, without realising that the bulk of it will be unnecessary, unwanted, inaccurate, and possibly even dangerous to our future lives and liberty. The age of the microchip can certainly give us unlimited data-storing potentialities; but we do not actually need it.

Already many people are functionally illiterate and bored with their lives. What concerns me, and I know it concerns many other people too, is that

'Of course, computers have given rise to such useful luxuries as the digital clock'

if we can control our purchasing from our armchairs, book arrangements without going out, check prices or whatever without the need to leave our homes, then we may create a future race that is soulless and even more insular, isolated and culturally dissatisfied than the one we have at present. Do you recall what I said on the subject of food technology on p. 32? We should interfere only when necessary, that was the idea – a principle of minimum interference.

This ought to provide an interesting way of looking at the future of the mega-computers. If we are going to install them, come what may, and then invent tasks for them to fulfill, we will be making a profound mistake. What we should do is the exact converse: we should observe what problems we really face, and introduce microelectronic devices to overcome those. That way we might find some of the unwanted drudgery taken out of our lives.

But if we ravage society with the machinations of countless superfluous machines that undertake tasks – like shopping – that are part and parcel of

our social intercourse, then we will put extra drudgery back in, and take out much of the fun.

WHAT USEFUL FUNCTIONS CAN MICROCHIPS PERFORM?

One useful area where the chip has done sterling service is by controlling factory-floor robots. The most extensive use of this principle is in the motor-car industry, where a host of menial and repetitive tasks can be given over to robots. Clearly, if robots could undertake dangerous and exacting tasks – like mining coal underground, for example – then the lot of the working population could possibly be eased.

Microchips in the home are already in daily use in calculators and clocks, in video games and microwave ovens, even regulating washing machines and air conditioning systems. The day of the automated house is already with us, for it was in 1982 that the first such home, built on the

39

outskirts of Phoenix, Arizona, was sold.

This futuristic dwelling was named Ahwatukee, an American Indian word meaning 'shining dream house', and it is controlled by five microprocessor computers. There are no keyholes at the doors. Instead, the would-be visitor presses a little button according to a pre-set code that can be changed at will. If the code entered is acceptable, the door opens. If it isn't, then a programmed synthesised 'verbal' message is given to the caller, which can be a strongly-worded request to go away if the occupant feels that way inclined.

All aspects of house security are integrated into this system, including fire alarms and burglar warnings. It could be used to regulate and monitor the exact levels of energy used in heating the house at night, or cooling it on hot summer days, and even to water house-plants when the soil was becoming dry.

The lights in the house are automated, so that they come on at dusk and optimise the interior light levels. There is even a message keyboard for casual callers. They can write their requests by punching them into the central control system, and a video screen can give them a time to call, or some other appropriate message.

There are two important features of this house that are worth emphasising. One is the cost; the entire computer system was installed for $30,000. The second is simplicity. When designing the system, a foolproof and easy-to-read mode of operation was the aim. You do not need to be a programmer to use the computer terminals, or even to have a knowledge of computerese. The systems are controlled with keyboard systems much like those installed at autobank outlets which dispense cash and give bank account statements on demand.

The future is rooted in cost efficiency. Labour charges have become prohibitively high, whilst computer costs have dropped. Between 1953 and 1980, for example, wages in Britain went up by eight times or more, whilst prices rose in the same period by around six times. In 1953, the catering company of Lyons, based in London, installed a computer to handle their accounts department. They reckoned it did the work of over a thousand full-time clerks. By 1980, the cost of that installation (£100,000 in 1953) had been – in real terms – reduced to one-fortieth of that sum. In economic terms, then, the electronic processor offers the capacity to undertake mundane tasks at a very low cost. It may be that the present euphoria is going to wane, and that the fashion for microchips-with-everything will fade in time, as all fashions do. But once we have established a valid perspective for the new microchip era of tomorrow, the most astonishing thought of all will be 'however did we manage without them?'

ARE WE NEARING THE DAY OF THE ARTIFICIAL MAN? IS MAN REALLY A MACHINE?

He certainly is not. I am assailed by people who analogise the human body to a construction made from girders and elastic, but the body is not machine-like and it would help all of us if the fact was generally appreciated. People point out that the heart, for example, goes on pumping 3,500 million times before it ceases operation and then draw comparisons with mechanical pumps. Of course an engineer could not construct a soft, flexible substitute for a human heart and expect it to last as long, for the simple reason that the heart you end up with is not the heart you start out with before birth. The muscle cells are in a state of constant flux, new materials are being taken in and waste ones expelled by the metabolic cycle of the heart muscle, and much of its structure is in a constant state of self-regulated renewal and repair.

Not only that, but if the heart is given an excess loading, like a failing pulmonary circulation, it will grow larger in an effort to compensate. If you overload a mechanical pump then it will very likely stop working. What engineering device would actually double its capacity from inside, in order to cope with demand?

There was a memorable line in a programme on British TV on 23 June 1961, when Tony Hancock performed his now-famous script 'The Blood Donor'. In one scene he was talking to a fellow-volunteer at the clinic, and said:

'Oh yes, it's advanced a lot, medical science . . .
I mean, look at the things they can do these
days. New blood, plastic bones, false teeth,
glasses, wigs – do you know there's some
people walking about with hardly anything
they started out with.'

Now, not far away from a quarter of a century
later, we have to face the fact that this view (which
was popular in the years around 1960) is still an
overstatement. The number of implantable pros-
theses is surprisingly limited, solely because man-
made substitutes – which have to rely on mechanical principles – lack this essential quality of
inner feedback. A great number of natural phenomena can be shown to be theoretically impossible,
when analysed by the mechanical principles of
science. There are birds that 'cannot' fly, and
insects that 'cannot' exist, according to our
mechanistic models. The sooner we give back to
biology the dignity of its own unique nature the
better. Man is not a machine. And no machine can
ever imitate man.

HOW DOES THE HUMAN EAR COMPARE WITH A MODERN SOUND SYSTEM?

Here is one quotable example of how human
performance puts mechanical imitation in the
shade. It is the functioning of the ear in detecting
sound. So often are we confronted with impressive public relations proclamations on the latest
innovations by acoustic engineers and hi-fi
specialists that it would be tempting to assume
that the out-dated human ear had been left far
behind. Not so. The delicacy by which our robust
sense of hearing seems to function is unmatched
by technology, and the way in which the brain
sorts out the information that the ear collects is an
even greater mystery.

The pitch of a note is measured by the number
of sound waves per second that reach the ear. This
is its frequency, and it is measured in Hertz units,
abbreviated to Hz, and named after Heinrich
Rudolf Hertz (1857–1894) the noted German

physicist who first detected what we would now
call radio waves in 1888. The lowest sound that
the ear detects as a note is around 50 Hz, and the
highest – which sounds more like a faint hiss than
a note – around 20,000 Hz. This very high
frequency can be heard by young ears, and is more
clearly heard by dogs (whose capacity to hear very
high frequencies considerably exceeds ours, hence
the 'silent' dog whistle); most adults cannot hear
notes above 15,000 Hz.

But what is so remarkable about the ear is its
capacity to detect very quiet notes. The sound is
picked up by the ear-drum, which is set into
vibration by the force of the sound vibrations in
the air as they strike its surface. Taking a note of
1,000 Hz, a somewhat high whine and the order of
frequency to which human hearing is most sensitive, we can determine exactly how much movement in the ear-drum is produced when the
minimum audible sound level is detected.

The amount of back and forward movement in
the drum under these conditions is calculated to
be around one thousand-millionth of a millimetre.
This is less than the diameter of the smallest atom.
The movement of the basilar membrane in the
inner ear, where the sound is detected, is believed
to be nearer a hundred-thousand-millionth part of
a centimetre, which is approximately the diameter
of an atomic nucleus. When you bear in mind that
the ear, like all structures in the body, is composed
of complex organic molecules that are in turn
made up of atoms, then the fact that this organ can
detect the result of movements on such a diminutive scale is imponderably magnificent. No synthetic substitute could approach such discrimination as the human ear.

What is to me the greatest achievement of
human hearing is the way in which sounds are
picked out by the ear, or rather, by the brain that
interprets the ear's input. A single voice can be
picked out from a babble of chatter, even if it is no
louder than the other competing sounds, because
of the particular auditory characteristics of that
specific sound pattern. What is even more remarkable is the manner in which the brain can
recognise order in chaos: a choir singing sends to
the ear such an immeasurably complex jumble of
out-of-sequence frequencies that it ought to

'. . . a single voice can be picked out from a babble of chatter . . .'

amount to a roar combined with a high-pitched hissing scream. On a cathode-ray oscilloscope, the eye would fail to see any order in the muddled trace of sound waves displayed from such a source of sound. The brain, however, is able to sort out the principal frequencies, so that you can tell which notes are being sung and by which sectors of the choir. It seems remarkable that if you play two notes together of, say, 250 Hz, the brain may receive them out of phase with each other, at the rate of 500 sound vibrations reaching the ear per second. This phenomenon is known as interference, and the spurious note of twice the frequency of the individual sources is quite easy to demonstrate in the laboratory. A second source of spurious sounds is when two notes are played of considerably different frequencies. If one note generator emits a note of 150 Hz, and another of 750 Hz, then there will be a cancelling-out effect where the two sets of pulses are working against each other – i.e. when they are out of phase – and a doubling-up when they are reinforcing each other – when they are in phase. The result of this is that, instead of the two notes originally produced, the ear is confused by hearing the intermediate

frequency of 750 − 150 = 600 Hz.

If the notes are nearer in frequency, say 500 and 505 Hz respectively, then this artificial set of pulses will occur at the rate of five per second. In this case the sound has a 'fluttering' quality corresponding to the difference in frequency of the two original notes. The effect is known as 'beats', and the frequency of the extraneous pulses is known as the beat frequency.

In practice, this should mean that the noise heard from a choir would be a cacophony of interference notes, spurious sounds and interwoven beat frequencies. To the eye, looking at the oscilloscope trace made by the noise, that is just what it looks like. But the ear and the brain manage to make sense of the jumbled input, and no synthetic analyser can approach this feat.

Starting with the quietest sound that can just be barely heard, and using that as the zero on a scale, it is possible to measure sound in terms of its loudness by the decibel scale, usually abbreviated to dB. Thus if you move from 0 dB to 10 dB, you have experienced an increase in sound volume of ten times. From 10 dB to 20 dB is an increase of a further tenfold – i.e. 100 times the original loudness. Thus a 30 dB increase corresponds to a 1000-fold increase in loudness, 40 dB to 10,000 times and so on. Some indication of this scale is given in the table that follows:

Representative Sound Levels

Faintest audible sound	0 dB
Quiet room	40 dB
Normal conversation	60 dB
Alarm bell ringing	80 dB
Industrial noise	100 dB
Jet aircraft overhead/pop concert	120 dB
Jet taking off	140 dB
High powered rifle (heard by firer)	160 dB

The sounds we would regard as annoying are such things as a noisy motor-cycle, a pneumatic road drill or a petrol power saw a few yards away, and these are in the range of 80–100 dB. Continuous exposure to this level of sound can eventually impair hearing. Even in a discotheque sound levels can be above 100 dB for prolonged periods, and in front of the speaker array at a live concert (which is far and away the best place to take in the flavour of the music) the levels can reach 120 dB. This produces a temporary deafness, and it may take three days before the individual is aware that his ears are back to normal. Taken to extremes, this form of battering could permanently restrict the ability of the ear to hear faint sounds. There is something to be said for a general slight reduction of sound levels at pop concerts, particularly since it is the subtleties of sound quality that impart much of the character to the music made by different groups, which the audience like to experience in the round. The tendency to generate as much power output as possible often exceeds the optimum limits of the amplifiers used, so that distortion and flattening occur. Sound quality of a precise and crisp kind would be infinitely preferable to a barrage of unresolvable sound, and if the volume level was within safe limits (even though still very loud by normal daily standards) then all music fans might gain some aural enjoyment at the same time as keeping our ears safe from long-term degeneration. A buzzing head is part of the ambience of a concert, of course – but it can still be kept within safe limits.

It has been calculated that the sound energy produced by a million people all talking at once would not amount to 20 watts, which is too little to light the headlamp on a car. Clearly, the ear is designed to respond to low sound levels, and it is interesting to note that there are muscles within the inner ear that can stretch the ear drum tighter and prevent it from responding to sound so sensitively. When in use, these muscles can produce a lowering in perceived sound by up to 30 dB, corresponding to a thousandfold decrease in sound energy. This is one of the protective mechanisms that the ear employs to prevent overloading; indeed, many people can contract and relax these muscles voluntarily. The threshold of actual pain, caused by noise, is 120 dB. It is worth remembering that this means that the ear can tolerate sounds (around 1,000 Hz) that are a billion times louder than the quietest whisper it can detect.

EXPLAIN THE MIRACLE OF THE ARTIFICIAL KIDNEY . . .

The fact that we can keep patients whose kidneys are diseased alive on artificial kidneys, as they are called, seems to go against what I have said of the special nature of living systems and their own characteristic complexity. But let me outline how the artificial kidney works in practice, and how the natural kidney functions.

The kidney is spoken of as a filter, which is a crude way of putting it. A filter holds back some constituents, whilst letting others through. What the kidney does is infinitely more complex, for it lets virtually everything through to begin with, and then takes back in whatever it actually needs for the maintenance of the body's chemistry. At the same time it exactly regulates the concentration of the tissue fluids and the blood, eliminating excess water as it does so. That is more than mere filtration.

Each day nearly 50 gallons of blood are processed by your kidneys, under normal circumstances, which means that the body's blood supply passes through the kidneys approximately twice every hour. This does not give a fair impression, however, for the rate at which the kidneys can handle unwanted components in the central blood vascular system is impressive. One vivid demonstration of this is the intravenous pyelogram, when a solution containing iodine is injected into a vein and its movement through the kidneys is then observed by X-ray examination.

The body needs only one ten-thousandth of a gram of iodine daily and the dose given into the vein for pyelography is perhaps three grams. This is hundreds of thousands of times more than can ever be used by the body, and so the surplus is excreted. As you watch the X-ray pictures, the concentration of the iodine in the kidneys can clearly be seen and, in a normally healthy individual, this gross overloading of the system has been entirely eliminated within 40 minutes or so. This is a highly efficient process, and one marvels at it in action.

The secret of the kidney's success lies in the minute structures known as Malpighian corpuscles, and named from their discoverer Marcello Malpighi (1628–1694) who was professor of medicine successively at Bologna, Pisa and Messina and who became physician to Pope Innocent XII in 1691 at the age of 62. Each of those bodies is a rounded structure one-fifth of a millimetre in diameter and filled with a tuft of fine capillary blood-vessels. The blood-vessel that feeds these capillaries is considerably larger than the vessel through which the blood drains, and so pressure builds up inside the capillary tuft.

As a result of this build-up of pressure, the constituents of the blood plasma are forced out through the capillary walls. Not only the unwanted constituents are expelled at this point; most of the chemical constituents come through, useful and otherwise. If the capillaries were spread out, their combined surface area in a typical adult would amount to $1\frac{1}{2}$ square metres. The expressed liquid then passes through a series of tubules, each one roughly 35–40 mm long, where the materials that are required by the body are reabsorbed into the blood-stream. The blood supply to these tubules comes directly from the capillaries that carried blood away from the Malpighian body, and so their normal blood contents becomes restored. The unwanted waste materials remain inside the tubules and pass through the collecting ducts down the ureter to the bladder as urine. It is said that a conservative estimate of the total length of all these tubules put together would be 75 km, and their total surface area, if they were opened out, would be almost 6 square metres. The packing of such a complex amount of plumbing into two kidneys each 11 × 6 × 3 cm and weighing about 140 gm is no mean feat of organisation.

What is more remarkable, however, is the self-regulated reabsorption of the materials that the body requires. One experiment showed that the kidneys would pass 83 litres of water through the capillary tuft/tubule system (known collectively as nephrons) and then reabsorb 82 litres in the same period; the remaining 1 litre being expelled as urine. The 90 gm of sucrose that was expelled was likewise reabsorbed in its entirety. All the salt in the blood plasma was passed through the system, and most of it was reabsorbed (just enough to supply the body's needs) whilst the remainder was expelled in the urine. The bulk of the urea was rejected, and all of the sulphate and creatinine – waste products from the body – were rejected too.

This amounts to a carefully regulated measuring and absorption system, controlled by the kidney in response to the bodily needs. When the few chemical compounds listed here are considered, the process looks refined enough; but there are innumerable substances handled by the nephrons every minute of our lives, and each one is regulated by the same carefully balanced control system. We have only the slightest idea of how it all works.

This brings me to the use of the artificial kidney. Here the patient's own kidneys have been damaged by disease or degeneration, and the waste products ordinarily released through the nephrons are retained instead by the bloodstream, with serious effects on the metabolism; if the condition is untreated, death is inevitable.

The principle of the artificial kidney is simple to understand. The patient's blood is fed from a main blood-vessel through a bath containing the chemical equivalent of blood plasma. The blood is separated from the solution by a membrane through which chemical molecules can pass, but which holds back complex colloidal materials such as proteins. The process is known as dialysis, and this kind of membrane is described as being semi-permeable. What happens is that any chemical present in the blood that is not in the dialysis solution will tend to pass from the blood through the semi-permeable membrane into the solution, since it is a physical property of all dissolved materials to move in such a way as to rectify unequal concentrations (osmosis). Conversely, if a substance present in the blood is present in the same concentration in the dialysis solution, then there will be no movement across the membrane. And if there is a high concentration of a substance in the solution that can diffuse through the membrane, then it will enter the patient's blood.

The artificial kidney contains a container for the blood, with a large surface area presented to the solution, and as the patient's blood passes through it is bathed in the dialysing liquid which is kept at 37°C (blood heat). Any waste materials present in the plasma diffuse out into the surrounding bath of liquid. The carefully-controlled composition of the solution is an exact match for what the blood-stream requires, of course, so all the useful materials remain in the blood where they belong.

A more recent development relies on the body's inbuilt semi-permeable membranes, rather than passing the blood through an artificial one. The intestines and the lining of the abdominal cavity are richly supplied with blood vessels, and the principle that has been used is to pass the dialysing solution through the patient's abdomen. It bathes all these blood vessels, allowing all the superfluous components to diffuse safely away, and does not require any external blood circulation.

Most recent has been an extension of this technique to allow the patient to move about during the manoeuvre. Since the lining of the abdomen is known as the peritoneum, the use of this membrane for purifying the blood is known as peritoneal dialysis. When the technique is applied to a patient who can move about, it rejoices in the term ambulatory peritoneal dialysis. Like every other term in science and

medicine, the complex-sounding words lose much of their mystery when we realise how mundane are the meanings they conceal.

It is often said that the problem with the artificial kidney is its size, for the traditional dialyser requires a small room full of apparatus to function. In my view, this avoids a more important issue. The point about the artificial kidney is not merely that it does not selectively reabsorb materials, as a kidney does, but merely relies on passive diffusion; more important is the much-overlooked fact that the principle on which the process relies is a perfectly-balanced dialysing solution.

This is no natural phenomenon. To make dialysing fluid requires a technical team who are capable of making accurate measurements so that they can reproduce the chemical milieu present in the healthy bloodstream, and balance the components so that the dialysis works properly. Starting from scratch, this would require years of investigation and development work, and a large, well-equipped laboratory. All this is part of the workings of the so-called artificial kidney.

This is why I feel we delude ourselves into

thinking that we are somehow competing with nature. That room filled with apparatus in which a patient can be dialysed is only part of the substitute kidney that the patient needs to survive. The other part is hidden – it is the laboratory complex that makes the dialysis fluid according to the body's own composition. Our own kidneys do not rely on outside agencies; they carry out all the measurements, adjust their demands, take back exactly what they require – against the natural osmotic tendency of the kidney tubules – and return to the body's mainstream circulation a regulated and purified blood supply.

All this they do within the space of a few spoonfuls of tissue, containing miles of tubing and a relatively large area of membrane surface, and half the time working against the osmotic pressures that govern living organisms. Yet the most thought we usually give to a kidney is when it is chopped up and served in a steak and kidney pie: an undignified fate for something so impressive.

WHAT IS BLOOD? WHY IS IT SALTY?

The blood is the means by which the cells of the body are kept together in a common environment which is nourished, warm and oxygenated. There are about 6 litres of blood in the body, and the heart pushes out about 100 ml at each stroke, adding up to the best part of 500 litres an hour – and that when resting. Animal life first became established in the sea, when living cells could exist bathed in a salty liquid which provided them with the mineral raw materials for life, and effortlessly carried away wastes. A blood supply is a means of keeping living cells (which evolved in the sea) in an environment that is still, essentially, sea-water.

The composition of sea-water and blood is similar. The proportion of mineral salts in blood is 0.9%, whilst in sea water it is over 1.0%, but that is today. It may well be that at the time animal life was evolving on land, 400 million years ago in the Cambrian period, the concentration of sea water had not quite reached its present-day level, and if that is the case then the 0.9% we see in human blood may well be more nearly the same as the concentration of salts in the sea when we first emerged from the oceans and crawled on land. Or at least, when our earliest ancestors did.

Blood is a tissue. The fact that it is a liquid is neither here nor there: it is not a liquid merely because it contains large amounts of water, but because the cells in blood are not held together as they are in other tissues. The proportion of water in blood is 80%, exactly the same as in the kidney, and approximately the same as in nervous tissue (like the brain) and muscle. Most of the body contains this apparently surprising amount of water, and even solid bone contains 30%.

Blood is approximately half and half cells and plasma (on average 45% cells, 55% plasma) and the overwhelming bulk of the cell fraction is made up of red corpuscles. In the laboratory we call these (officially) erythrocytes or (colloquially) red cells. 'Erythrocytes' merely means 'red cells' in Greek. Red cells are not red, but a pale straw colour when looked at in the living state. In large numbers they appear red, because of their haemoglobin content.

Haemoglobin is a protein that is able to pick up large amounts of dissolved gases, like oxygen, and then release it again when required to do so. In the lungs the oxygen from the air readily combines with the haemoglobin in the red cells, and in areas of the body where the demand is high, the oxygen separates from the red cells and diffuses into the surrounding tissues. Without haemoglobin, blood would still carry a good deal of oxygen, but if you held your breath for more than five seconds you would run out of this vital dissolved gas. Haemoglobin is the reason why you can hold your breath for much longer, in some cases for several minutes.

Red cells are always spoken of as though they were rather like draughtsmen, for their shape is biconcave and rounded, like a sweet pastille. But that is how they look in bulk, and how they seem if you examine them when dehydrated and mounted for the microscope. In life – though they normally have this general outline – they are soft and flexible, rather like thin-walled bags of jelly. Red cells are capable of enormous amounts of distortion as they squeeze through gaps and narrow passages in the blood capillaries, and in some instances they can become elongated and sausage-shaped as they are forced along a narrow channel. Once free of restriction, they gently recover their familiar shape, often oscillating in and out like a polythene bag full of water.

There are roughly five million red cells in a cubic millimetre of blood, amounting to 35,000,000,000,000 in the entire circulation of an adult human.

The white cells are a varied population, essentially concerned with the fight against disease and resistance of the body to outside agencies – foreign tissue, for example. There is on average one white cell or leucocyte for every 625 red cells, giving a 'count' of approximately 8,000 per cubic millimetre, a whole body population of 56,000,000,000. There are also tiny bodies, smaller than these cells, called platelets. There are 250,000 of these per cubic millimetre, and they are concerned with the clotting of the blood when a vessel wall is damaged. There are thus some 36,806,000,000,000 separate bodies floating in your bloodstream at any given time, or more than twelve thousand times the human population of the globe.

White cells have a life-span of about three weeks, whilst red cells survive for 15 weeks. Most of the cells are formed in the marrow, some of the white cells in the spleen and the lymph nodes. Once they are fully formed and are released into the blood-stream, the blood cells do not normally reproduce or undergo division. They live out their mature lives and then die and are scavenged by other cells in the body so that their vital constituents (particularly the haemoglobin of the red cells) can be recycled and used in the production of the next batch of new cells. Many of the white cells are much like a simple amoeba, living in a pond, a reminder of the fact that we consist of great colonies of independent cells. If you observe a white cell scavenging around, moving, sensing, feeding, it is hard to imagine that you are watching anything as 'lowly' as a single, lone cell. I have no doubt that the mechanisms that go on inside single cells are very much more discriminating and complex than science admits. Meanwhile the sight of a single white blood cell

moving about in the salty bath of the blood plasma takes us right back to the earliest single-celled animals that used to inhabit the primaeval seas. The view has not changed much in all those five hundred million years.

WHAT ARE BLOOD GROUPS?

These are categories of blood types that occur in all the races of mankind, some of which interact with each other and cause clumping of the red cells with repercussions for blood transfusion patients. There are four major blood groups known as O, A, B and AB. The reason for this is as follows. Within the human populations, there are found two different components in blood that can cause the cells to clump together or agglutinate. Because they generate agglutination, these components are referred to as agglutinogens, and are designated A and B respectively.

People who have neither type of agglutinogen in their blood are said to be of Group O. A minority have both types, so they are designated Group AB. The people who have one agglutinogen or the other are known as Groups A and B. The older convention for the blood groups was I, II, III and IV; but the AB system is far better as it indicates exactly what is the agglutinogen status of a given individual.

A second blood group factor is the so-called Rhesus Factor, abbreviated to Rh. It was discovered in 1940, when it was found that cells from Rhesus monkeys injected into guinea pigs stimulated the formation of antibody potentially lethal to the blood cells. It has since been found that a similar condition occurs in humans. If a person bearing this factor donates blood to a person without it, the recipient produces antibodies which, at the time, pass unnoticed. However the next time that the same blood is administered to the patient, the antibodies that are, as it were, 'lying in wait' attack the blood cells and a severe (even fatal) reaction ensues. Though this happens after blood transfusions, it also affects new-born babies when the father has bequeathed to the foetus the factor which is missing from the mother's blood. The mother produces an anti-Rhesus factor which attacks the blood cells of the baby, and if undetected only urgent total transfusion with blood of the right group can save the child's life.

Blood which contains the factor is known as Rh-positive, blood that lacks it Rh-negative. White races are notable for their tendency to suffer the effects of Rhesus incompatibility. Among American Indians, pure Negroes, Chinese and Japanese, less than one per cent are Rh-negative. White races have fifteen per cent in this category, though no one knows why.

There are therefore three factors of note in the blood groups of mankind: agglutinogens A and B and the Rhesus factor. Between them they make up a total of eight different groups, A+ and A−, B+ and B−, AB+ and AB−, O+ and O−. There are some fifteen or so other blood group systems of minor importance. These include the Lutheran, Kidd, Duffy, Lewis, Kell, MN, P and Xg groups, and they are mainly of interest to geneticists. The Xg group is of particular interest since it is sex-linked, and the gene coding for it occurs on the (female) X-chromosome.

The commonest group is Group O, with 44% of the population. Group A amounts to 41%, Group B 10% and Group AB, 5%. Since Rh-negative individuals amount to only 15% of the population, it follows that the rarest of these blood groups would be AB−, the commonest O+.

WHAT IS THE BEST WAY TO REMOVE BLOOD STAINS SCIENTIFICALLY?

If you apply water to fresh blood stains on an item of clothing (where you have cut yourself shaving, let us say) then the red cells burst and their colour disappears almost immediately. A few drops of cold water will do the trick. Failing that, a little saliva might do. If the stains have dried on then pre-wash the garment in a biological detergent. This kind of cleansing agent contains enzymes extracted from cultures of microbes and the effect of the enzymes is to digest the blood stain out of

the cloth (pp. 23–4). Beware of using enzyme preparations on woollen materials. If they are left in contact for too long the wool will start to be digested too.

WHAT ARE DISINFECTANTS MADE OF?

The term 'disinfectant' means an agent that eliminates a source of infection, strictly speaking, so that penicillin and paraffin flame-throwers come into the category. In current use, though, the term has come to mean any liquid preparation that has a characteristic smell, and which kills microbes. The smell is the distinctive part of familiar antiseptics.

The original disinfectant was carbolic acid, used in a spray by Lister and others to control air-borne infections. Unfortunately it soon became apparent that you had to use equally strenuous methods to control the effects of the disinfectant: it is a caustic substance and eats into unprotected tissues. Carbolic acid is known in science as phenol, and a later disinfectant from the same stable was lysol. It is still used today to treat drains and spillages of potentially infectious materials, but it is not suitable for use in personal hygiene. Some of the brown liquid disinfectants are safe to use in contact with the body, including chlorxylenol which is the active ingredient in Dettol*, and there are less colourful stable-mates including trichlorphenol, sold in TCP*.

Added, in the recommended dilutions, to stale waste water they are effective in preventing the further growth of organisms of decay and so in cutting down odours. There is actually very little danger of infection from drains. If someone suffering from cholera has heavily contaminated a drain you are cleaning then you may well catch it too, but diseases do not generate themselves spontaneously in malodorous situations. The germs have to get there from somewhere. The scent of pine or the aroma of phenols is comforting to our senses, and makes us feel that someone has been busily caring for us and attending to our welfare, but the real effect on the spread of disease is minimal in Western society.

It is interesting to recall that bad smells have traditionally been associated with the onset of infectious diseases. A sweet-scented nosegay was carried in the Middle Ages to ward off disease germs, and the 'pocket full of posies' in the nursery rhyme alludes to a bunch of flowers that would have been carried to ward off the plague. The 'ring a ring of roses' refers to the rash patients developed, and the 'tishoo, a-tishoo, we all fall down' of the chorus graphically describes what happened if you were a victim. Sneezing was one of the first symptoms of the plague (which is why we say 'Bless you' to a sneezer), and I suppose you could say that falling down – dead – was the last. Malaria, in much the same way, means 'bad air', and to this day you can find people who tell you that a foul smell will make you ill.

There is more than most people would admit in this age-old association between odours and illness. If you look at the advertisements for household disinfectants they often stress the smell of the preparation, rather than its value as an antimicrobial solution. We are cajoled into buying a disinfectant because it smells like pine-

woods, or fresh lemons, or just because it smells strong. But note that it is the idea of the 'good smell triumphing over bad', and the promise of the powerful disinfectant killing our disease-causing enemies, the germs, that lies behind it all.

One of the best household products for killing microbial life, however, is not sold as a disinfectant at all, but as a bleach. The household bleach is a solution of sodium hypochlorite, $NaOCl$, and it reacts with any organic substance to give off chlorine. A small amount of hypochlorite in a large pail of contaminated water will kill any bacteria almost instantly. So the use of bleach in keeping a lavatory hygienic is ideal: at the same time as bleaching the unit and making it look clean and hygienic, you are killing off any bacteria that may be left around.

But there is a word of warning that should be emphasized. Hypochlorites react with any organic material, not just with germs. Suppose there is a lot of waste matter in the area you are trying to treat. All that will happen is that the $NaOCl$ is exhausted in reacting with the waste. In consequence, the chlorine is rapidly released in contact with the organic substrate, and that can produce a dense and choking atmosphere which can make anyone ill (chlorine produced in such a manner has been used as a poison gas, for instance in the first World War). Naturally, once the hypochlorite has been exhausted in this way it is no longer effective against germs, so anything lurking within the waste material will survive.

It is therefore important to realise that household bleach should be used at the recommended concentration, and not too strong. It should also be used in situations where the container is almost entirely clean and free of surplus organic matter. It will then work powerfully in removing the last few traces of waste material and in sterilising the entire area in which the bleach is contained.

Hypochlorite bleaches are without doubt the most useful means of sterilising food containers, and are particularly beneficial for the home brewer. If you use sulphur dioxide tablets then there is always residual sulphur dioxide lurking around, and the quantity of sulphur dioxide produced depends on how many tablets you use.

* Trade mark.

With hypochlorite bleaches, the chlorine is only released in contact with actual organic matter and if there is no contamination with this kind of material there will be next to no chlorine liberated from the bleach solution.

To sterilise wine bottles or fermentation vessels using this method, wash and clean them as well as you can and then rinse in clear water. Add a little bleach to each container and top up with tap water (the proportions will have to be calculated from the dilution recommended on the bottle in which the bleach is purchased; otherwise use 5 ml of the bleach per half-litre of water). Leave that to stand for a few minutes, then empty out and rinse with tap water once again. Do not fear that the water will contaminate the bottles, as tap water is sterile to all intents and purposes. Keep a bowl of hypochlorite solution handy, and then it will be valuable for on-the-spot sterilization of anything that might become contaminated. If the end of a syphon tube brushes against the floor, for example, rinse it quickly under the tap and leave it in the bleach solution for half a minute or so. Any contaminating organism will be eliminated by this treatment.

CAN HYPOCHLORITE BLEACHES BE USED ON MY HAIR?

Only if you do not object to your hair turning white, brittle, and falling out in broken handfuls. The kind of bleach used on hair is hydrogen peroxide, H_2O_2, which is similar in composition to water, but with an extra oxygen atom. Hydrogen peroxide reacts in contact with organic matter to form water and oxygen and it is this reaction that produces the bleaching effect:

$$H-O-O-H \rightarrow H-O-H+O$$
$$H_2O_2 \rightarrow H_2O+O$$
Hydrogen
peroxide $\rightarrow$ water + oxygen

You could call water 'hydrogen oxide' and the hydrogen peroxide could then be known as 'hydrogen dioxide' quite correctly. The amount of peroxide bleach that has to be used on human hair is a delicate matter and home bleaches contain a range of stabilisers and conditioners to help minimise hair damage. Professionals, who are used to such things, frequently buy peroxide in large jars and then dilute it for use themselves, but I would not recommend that to anyone outside the snip-and-trim business.

Hydrogen peroxide decomposes to oxygen and water very easily, even on contact with daylight, and it cannot be bought in the pure form. What you purchase instead is a solution of peroxide in water. The strength is not indicated by the percentage of the solution that is peroxide, but by a time-honoured formula that describes the volume of oxygen the solution would give off. A litre of 20-volume peroxide contains the equivalent of 20 litres of oxygen, in other words. The professional hairdresser uses 20-vol or 10-vol peroxide, carefully diluted down.

A more modern equivalent is a preparation that can be sprayed onto the hair and left on whilst the user spends time in the sun. The action of the heat from the sun's rays causes the bleaching to take effect where the rays strike, and so the result is a lifelike 'sun-bleached' look. But do not be fooled. Bleached hair is brittle hair, and the use of conditioners can only keep the brittleness at bay; it cannot eliminate it altogether. Too frequent use of bleaches can seriously damage hair, and so moderation is the keyword.

YOU HAVE BEEN QUOTED AS SAYING YOU KNOW A CURE FOR BALDNESS. IS THIS TRUE?

Yes.

WHAT IS THE CURE?

Unfortunately it is the mutilating operation known as castration – not to be recommended for my readers! It is true that castration is a

GELD & SPAY
HAIR
CONSERVATION
CLINIC

percentage has never been documented, for there are too few eunuchs in the world to make such a figure meaningful, but it is known that a proportion of eunuchs retain the ability to perform sexually.

Other forms of baldness treatment are uncertain. Punch grafts are reasonably successful, but they are sometimes unsightly and they are not permanent. The most recent surgical method involves removing a tapering segment of the hairless scalp and drawing the edges together, stretching them across the skull. In many instances the scar edge is reportedly unsightly, and a perfect repair is almost impossible. But it is a clever idea, as all obvious proposals are, and may yet prove to be the safest and most reliable amelioration of this widespread and much-resented condition.

Hair is a remarkable substance. In humans the main growth is restricted to distinct regions of the body, although of course we have hair all over our bodies, which reminds us of our furrier ancestors. The insulating effects of fur can be exemplified by the husky dog that can sleep in the open air in a temperature of $-40°C$ without losing any measurable amount of heat, and desert sheep that can register a temperature in direct sunlight near $90°C$ whilst remaining cool underneath their blanket of fleece. Hair itself is made of keratin, which is a poor conductor of heat; and the layer of air that is trapped between the strands is also an excellent insulator. It is because of this fact that the fibre-packed 'continental quilt' or duvet is so successful at conserving the body's warmth on a cold night.

preventative against baldness, since it is known that eunuchs do not become bald. Some baldies have turned this argument on its head, and claimed that if eunuchs are immune to hair loss, then the more hair you lose naturally the more virile you must be. There is no suggestion that this is the case, and it is more likely that bald men who say this are trying to cover their feelings of insecurity. As far as one can tell, there is not the least relationship between sexiness and baldness. It should be said, however, that many women find bald men peculiarly attractive, and there are plenty of artificially bald celebrities who find a gleaming pate nothing but a boon.

Castration may prevent baldness, but might it also be a useful cure for rapists, sexual perverts and social deviants? In recent years many spokesmen have called for mandatory castration of criminals such as rapists. However much this reveals about the sadism of the ostensibly innocent members of the community, it does not accord with scientific facts. Castration does not preventual sexual urges in everyone. The exact

DO STRING VESTS KEEP YOU WARM?

Not as well as people claim. It is true that air is a very effective insulator (see previous answer) and that air trapped in the fibres of clothes can keep the wearer warm. It was from this that the idea of the string vest evolved, as well as being economical to make and quick to assemble from raw materials. The number of machine movements to make a string vest are far fewer than to construct conventionally woven textiles, which adds up to a manufacturer's saving of considerable proportions.

The difficulty lies in the size of the pockets in which the air is trapped. If the air was held in a firm blanket it would indeed insulate. But the open structure of a string vest means that movements of the shirt and outer garments can flush out a considerable proportion of the (warmed) air in the pockets of the vest, and replace it with displaced fresh air which is cold. The point about air as an insulator is that it must be trapped in position. When it can move in and out, as it can in this case, it carries the bulk of the heat with it.

PEOPLE SAY THAT COLD DRINKS, RICH IN CALORIES, ARE JUST AS WARMING AS HOT ONES. CAN YOU EXPLAIN THIS?

No, I cannot, for although this is very widely taught it is wrong. The body has to burn up energy-rich materials in the liver and elsewhere in order to maintain the temperature of the circulating bloodstream at 37°C. Calorie-rich foods are an important component of the diet of anyone who needs to resist the cold.

But although this is how metabolic energy is obtained, what people forget too easily is that any heat taken into the body – even if it is in the form of stored heat in a warm drink – is an energy input. A Calorie (see p. 22) is the amount of energy you need to raise a litre of water by 1°C. A warm drink, at 45°C, would contain 40 Calories more than a cold drink inbibed at 5°C. Put another way, if that amount of warmth is taken directly into the body, then a correspondingly reduced amount of the body's food reserves will have to be consumed. Not only that, but the psychological effect of mulling over a warm drink is a great bonus in restoring anyone who feels oppressed by the cold weather. So the idea that it is only the Calories in the food you consume which make you warm is not only wrong, it is also misguided in principle.

CAN PEOPLE TRULY WALK ON FIRE?

The legends of fire-walking stretch back to antiquity, and there are several areas of the world today where the practice goes on. Ceremonies in Sri Lanka and Southern India, the South Pacific and even in the Balkans are still witnessed today. A similar episode involving the young King Edward VII and a cauldron of molten lead was reconstructed earlier this century, though no explanation for it was finally agreed.

Fire-walking still takes place in the coastal settlements of Sri Lanka north of the capital Colombo. A pit by an ancient Hindu shrine is filled with coals so hot that the 'stokers' regularly sprinkle themselves with water to cool their skin. The first to walk across is the priest of the cult, who almost strolls across the coals without apparent mishap. He is followed by the faithful, young and old (and some of them carrying babies in their arms) who walk across the glowing bed. Similar ceremonies take place across the straits in India. Travellers who have witnessed the ceremony say that it is impossible to approach the fire-pit, which may be 15 ft long, closer than 30 ft. The radiant heat cannot be endured at closer range.

In the South Pacific, fire-walking ceremonies centre on a circular pit up to 20 ft across and 4 ft in depth, filled with burning logs and volcanic rocks. The priest, known as the Mbete, tests that the logs have all burned away and the stones are ready by scattering handfuls of leaves across the surface. They burst into flame at once. Then, at a signal, Mbete leads the volunteers across the pit, sometimes walking in a circular path around the glowing embers, before emerging back onto the surrounding earth unscathed.

The Balkan version is now very rare, but I have witnessed it in a mountain village north of the Black Sea coastal town of Burgas. Here, conical fires made from long logs are lit and burn for hours whilst feasting and drinking go on as the sun sets. Eventually, at midnight, the fires are raked out into large flat beds of glowing charcoal perhaps 25 ft across. The men (who all belong to the same family and have practised the same art for many generations) remove their shoes or sandals and take some cautious steps across the edge of the fire-bed. The path they take becomes increasingly bold until groups of people are walking across the fire in a kind of ritual dance.

In Britain, the story about the young King Edward VII tells how he was sent to Edinburgh to be taught by the scientist and politician, Lyon Playfair (1819–1898). Edward, then Prince of Wales, was taught by Playfair how to plunge his hand into a cauldron of boiling lead without sustaining any injury. The story soon became part of the royal family folklore and has been passed down to the present day.

Doubtless the story has been embroidered with the passage of years, but in 1938 one of Playfair's successors, James Kendall, decided to restage the demonstration. In his version (which is probably nearer the truth) molten lead was poured from a crucible in a stream through which he – and later his daughter – passed their hands, fingers apart, without injury. Commentators on the experiment admit, somewhat ruefully, that the explanation is still not clear.

My view is that the answer lies in two specific areas. One of these consists of limiting the amount of heat that is brought into contact with the body: this can be done by cutting down the heated area that comes into contact with the skin, by keeping the time of contact down to a minimum, or by the use of materials that – even if they are actually at high temperature – are poor conductors and will not bring to the surface excessive amounts of stored energy from within the heated mass.

The other area concerns the high specific heat of water (p. 4) and its high latent heat of evaporation, that is the amount of heat energy it takes to change water from the liquid to the vapour state. All of these factors are interrelated in the fire-walking episodes described above, and when they are overlaid by religious or meta-physical trance-like states, or similar conditions of detachment brought about by the consumption of alcohol, then the examples we have considered become easier to understand.

To save space and to make the discussion easier, I will assign shorthand symbols to the factors listed above, as follows:

CATEGORY 1: Limiting heat input
    1a – Reducing contact area
    1b – Reducing contact time
    1c – Using poor conductors
CATEGORY 2: Relying on evaporation of water
    from skin.

First, the examples from India, Sri Lanka and Fiji. The participants rely on walking on glowing coals that have been stoked by men who cool themselves with water. These attendants clearly rely on 2 to keep themselves comfortable. The fire-walkers rely on 1a, because they are supported on the uppermost edges of the cinders, which are also the coolest parts of the fire. The priest himself seems oblivious to 1b, perhaps because he is a practised participant and has developed thickened soles to his feet as some form of protection, and also because of his religious trance-like state. His followers walk faster than he does, so they appear to be utilising 1b. All of them rely on 2, because it is the water content of the soles of the feet that absorbs much of the heat energy, as the skin – which is a poor conductor – loses moisture as vapour.

'. . . his followers go through faster . . .'

In Fiji the rocks are spongy and volcanic in origin. This means that the contact area is small (1a) and the capacity of these rocks to hold heat is

53

limited. They are not efficient conductors, so heat lost from the outer layers is not rapidly restored by heat from the centre (1c).

In all of these ceremonies, travellers speak of the great heat waves that seem to strike them. But that is because the witnesses are standing there, facing the fire, waiting for the ceremony to start. The stokers who douse themselves with cold water are, similarly, close to the fire and facing its considerable output of radiant heat for a prolonged period. In this way, the skin's natural heat-resisting properties are partly exhausted. The poor conductivity of the skin is overcome because of the continued bombardment with infra-red radiation, and the moisture content of the skin is depleted because of the prolonged time of exposure. The participants, on the other hand, are fresh to the fire. They come with cool skin which is replete with a good supply of natural moisture, and this provides the 'heat sink', as a physicist would say, to dispose of the input.

The Balkan fire-walkers do not stand on heated rocks, but on charcoal embers. Here too the skin is well supplied with moisture, for they do not remove their socks until immediately before walking, and this prevents the moisture from evaporating from the soles of the feet which keeps the water content near the maximum. The charcoal is light and open in structure, which means that it does not store much heat energy, even when its temperature is high; and the area of contact is limited because of the porous and fragmentary nature of broken charcoal. Thus there is a small measure of factors 1a, 1b and 1c, coupled with 2.

The story of the molten lead cannot rely on 1c, since lead is a good conductor and heat lost to the skin from the surface would be rapidly made up from within the lead mass. But the pouring of lead means that as it passes across the fingers the contact time (1b) is kept to a minimum. I also believe that there is a special extra mechanism at work in this example which potentiates factor 1a.

When the lead strikes the skin, some of the heat transmitted will be absorbed in vaporising water from the skin surface. The heat will not readily be transmitted to the inner layers of the fingers, because skin is a poor conductor of heat, and for short spells this vaporisation will act as a protective.

But what happens as the vapour escapes from the skin itself? I believe it forms a protective layer that actually prevents the molten lead from coming into direct contact with the skin surface for part of the time. My theory is that as the hot lead strikes the fingers, if enough heat is transferred to start the water vaporising, then the escaping gas will actually keep the rest of the lead away from the skin. The stream is, as you might say, blown away by the vapour that the heat of the lead itself is generating.

These explanations do not mean that anyone can go out and try fire-walking. The exact methods of controlling the many factors, particularly those in Category 1, have been worked out by generations of practitioners. In addition, the right state of mind – whether it is induced by fervent religious belief or equally assiduous consumption of plum brandy – is also an important factor.

Although these answers to the mystifying phenomenon of fire-walking remove some of the incredulity that many people experience when faced with the fact, it remains true that the ceremony is dangerous and, for the participants, it can be damaging. Fire-walkers in the religious ceremonies do not talk about their experiences because they are associated with patterns of worship. And their Balkan counterparts, who are sometimes more communicative, exhibit calloused, discoloured, and sometimes burned and blistered soles to their feet. So fire-walking, even if it can be partly explained by science, remains an impressive and awesome phenomenon.

DOES THE PREVIOUS ANSWER EXPLAIN WHY I CAN PASS MY HANDS THROUGH THE STEAM JET OF AN ESPRESSO COFFEE MAKER WITHOUT BURNING MYSELF?

No, it does not. What happens here is unrelated to the previous examination of fire-walking. When a gas expands suddenly, it loses heat energy in the process, an effect known as adiabatic expansion. So if superheated steam (often well above boiling point, 100°C) is released from a jet, it cools dramatically as it expands. By the time it is several centimetres from the opening of the jet, it can be only warm to the touch, and further away still the jet may even feel cold.

But care should be taken in playing with a steam jet. Before the steam has had adequate time to expand it will still be above boiling point. Exposing yourself to that onslaught would be enough to damage the skin seriously and cause blistering. Even the protective value of the skin's moisture content in evaporation cannot assist here, as the hot jet itself is saturated with moisture and so no significant further evaporation from the hand would occur.

WHY DO I BLOW ON MY FOOD TO COOL IT, YET BLOW ON MY HANDS TO WARM THEM?

The only time I was presented with an answer to this question it was as follows: 'It must be because in each case you are trying to change something from hot or cold to body heat. If you hands are *below* 37°C, then the breath will heat them up to that level, whereas if your food is well *above* that temperature, your breath will cool it down towards 37°C.'

I like an answer like that. It is thoughtful and inventive. The fact that it is also wrong does not make me appreciate it any the less.

Perhaps as you read this section you would care to demonstrate what actually happens. Open the mouth (comfortably, not excessively, for I would not want onlookers to embarrass you by their response) and gently breath out onto the back of your hand. Hold the hand close to the mouth, about an inch away from your lips, and note how that feels. The breath is warm, and the back of your hand is heated by it. Note too how when you remove the hand and it moves about again (perhaps when you turned the page) it feels cool. This is because of the moisture from the breath evaporating on contact with fresh air, and the

energy it consumes in the process has to come from somewhere.

Part of that heating came from the mere fact that your breath is exhaled at 30–32°C (note, not at 37°C) and the surface of the back of your hand is lower than this. But a vitally important mechanism which reinforces the effect lies in the latent heat of vaporisation of water (p. 4). I have explained how, when a substance moves from liquid to vapour, it absorbs a considerable amount of heat energy. In the case of water this is a larger-than-average amount.

When the converse effect occurs, i.e. when a vapour changes to a liquid, then the heat energy is given out once more. So if moisture is condensing onto the back of your hand in the previous experiment – and it will indeed do so because the breath is laden with water vapour, and your hand is cooler than the breath – then as the condensation takes place the stored latent heat in the water will be liberated. This is another argument against playing with steam jets (p. 54) since if hot water vapour condenses onto your hand, the released latent heat will intensify the effects and could cause a serious scald.

So that explains why it is that you can breathe on something to warm it up. We now have to understand how the breath can equally be used to cool things down. For this I would invite you to carry out another experiment. Hold the back of your hand to your mouth once again, this time further away (say ten centimetres). Close the lips to form a very fine jet, and direct air by breathing out onto the back of your hand. Note that, this time, the air feels very much colder.

One of the reasons for this is the process of adiabatic expansion which we considered earlier on this page. As the air is compressed in our mouth, and then released by blowing, it expands and loses heat energy. In this way the air at around 30°C can cool, perhaps to 15°C.

The second reason is concerned with latent heat once more, for as the jet of cool air strikes your hand it takes water from the skin, and the evaporation of this moisture to water vapour removes further heat energy. The two processes mean that blowing on your hand makes it cold, whilst breathing, by contrast, warms it up.

It is now simple to account for the apparently contradictory observations posed by the question. You blow onto your food to cool it partly because the adiabatic expansion of air through pursed lips generates a stream of cooled air, which carries away heat from the food; and also because water evaporating from the food into the air stream causes the consumption of still more heat energy. If you were to blow on something hot and dry, like a poker from the fire, then the cool air would carry away the heat but there would be no further loss due to latent heat. So blowing onto a damp substance cools it more effectively than blowing onto something that is dry.

The reason you breathe onto your hands in the winter is because the hot breath warms them, and the condensing moisture on your chilled skin liberates latent heat. You will probably have noticed that it is instictive, after breathing on cold hands, to put them at once into your pockets, inside a coat, or back into gloves. This instinct arises because, if you were to leave the newly-warmed skin in the fresh air, the moisture would evaporate again, taking with it heat energy. By plunging your hands into a protective enviroment this heat loss is avoided. What is happening is that previous experience has been subconsciously learned, and you carry out – without even thinking about it – a series of actions that apply the theories of physics in a highly efficient manner.

IS PHYSICAL MOVEMENT THE BEST WAY TO KEEP WARM, OR WOULD A STIFF WHISKY BE A BETTER IDEA?

I remember once at school when the dormitory was unspeakably cold and the blankets provided by the establishment were hardly enough to warm a turbot. A protest was lodged with the house-master, who turned out to have been a keep-fit fanatic and an expert in amateur gymnastics. He stood in the dimly-lit door and bellowed at us: 'Physical movement is what you want, laddies, physical movement! So don't just lie there shiver-ing – DO SOMETHING. Come on there, wriggle about a bit!'

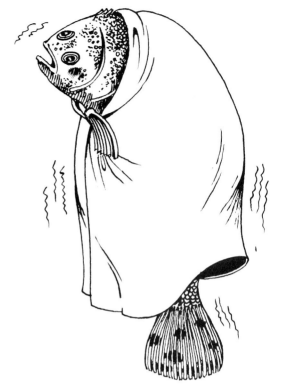

'. . . blankets hardly enough to warm a turbot . . .'

Heedless of the fact that this rejoinder for rhythmic activity beneath the bedsheets might be misconstrued, we all moved about in our cold beds as best we could whilst the instructor returned to his own room, doubtless well in-sulated and deep in swansdown.

In those circumstances, his advice was worse than useless. What little heat had been trapped in our beds was quickly washed out by the waves of fresh, cold air that each movement admitted. Within a few minutes we were colder than ever.

Physical exertion in such a situation cannot solve the problem. The liver burns stored food reserves and is always one or two degrees above the rest of the body. It is here that the central heating system of the body resides (p. 19). The muscles of the body also liberate heat when they work, but wriggling about is insufficient to make a useful difference. The only way in which our ill-inspired gym instructor could have applied physi-cal exertion to getting us warm would have been to get everyone dressed and running around the playing fields for half an hour. By the time that was over, the body would feel extra warm, since the over-heating of the muscles had been counterbalanced by the dilation of the blood vessels in the skin, and – since it is in the skin that the temperature-sensing nerves reside – the whole body would have felt glowing and fit.

However, knowing how such a project would have been greeted by all of us at the time, I am glad that the scheme did not occur to him.

Alcohol can produce a warm feeling, too, but it is an illusory sensation since the body itself is not generating additional heat, but is losing it. The effects of alcohol are many, but one of the physical results is a dilation of the blood vessels in the skin (of the same kind as we have considered above). As a result of this, the warm blood-stream is directed into the skin and the body surface becomes flushed and warm. An outside observer would say you looked pink, rosy, glowing. You would say you felt warm, cosy, relaxed. But what has happened is more serious.

The direction of extra supplies of blood to the peripheral blood vessels causes an increased rate of heat loss from the body, so you are – in terms of energy stores – running out of heat. Alcohol succeeds in cheating the senses and causing further heat losses at a time when they should be avoided.

The best way of keeping warm is to retain the heat that the liver, together with the remainder of the body, is always producing. Thermal or insulating clothes are ideal. A fleece underblanket, topped with your normal bedsheet, will seem so effective as an insulator that you might be forgiven for thinking that an electric blanket had been installed. Thick socks and gloves for those with a tendency to develop cold extremities are invaluable. For hypothermia victims brought in from distant locations under bad weather conditions, an insulated blanket is provided fitted with a reflecting sheet that looks like aluminium cooking foil. The reflecting layer keeps in heat that would otherwise radiate outwards, so helping to conserve as much body warmth as possible.

Clearly a warm room is the most convenient answer to a cold body. But many people have to go outside, and many of those who have to stay indoors cannot readily afford to heat rooms adequately. This applies particularly to the elderly, who are often the least well-off sector of society as well as being the most at risk from cold. If a warm room is to be the answer, then national governments should make a free allocation of fuel available to everyone, with units above the allocation being metered and paid for as at present. This would allow poorer people to use a measure of heating without incurring penalties, whilst giving those who wanted to use more energy a bill that was probably much the same as they had previously had. I have been told that a system such as this is already in use in Ireland, though I know nothing of its details, but it is such an obviously beneficial ideal to supply everyone with some warmth as an entitlement, that I would hope other nations would follow suit as soon as they can.

CAN THUNDER TURN THE MILK SOUR?

This old wives' tale still persists in many areas. It is untrue, since thunder is due to the explosive effects of suddenly heating a column of air during the discharge of a build-up of static electricity in the form we know as lightning. Milk turns sour because bacteria of the genus *Lactobacillus* grow in the liquid, obtaining their energy by oxidising the sugar, lactose, into the sour-tasting lactic acid. Electrical discharges, and the noise resulting from them, are unlikely to influence the growth rate of bacteria in milk.

I can only conclude that the story arose because the circumstances in which milk will turn sour are when it is kept warm, since lactobacilli grow slowly in the cold. Thunder occurs most frequently in warm weather, which would relate the two phenomena. Furthermore, thunderstorms occur when columns of heated air rise from the ground into the cooler upper atmosphere, and hence they are most frequent at the end of a hot day. This fact seems to have escaped attention, but in my view it makes it very likely that thunder and sour milk might be superstitiously connected. It would take all day for a bottle of milk to turn sour, by which time a thunderstorm – if one was due – would usually occur. The timing would be perfect for someone to drink milk in the early afternoon and find it perfectly acceptable, then watch the late-afternoon thunderstorm (which would distract attention from milk-drinking and a host of other mundane activities) only to find that – by the time the next drink of milk was taken – the taste had turned and the milk was soured.

The timing of both processes is so close that it would be perfectly possible for people to conclude that the thunder had affected the milk.

DO OSTRICHES BURY THEIR HEADS IN THE SAND?

It would be reassuring if they did, since the metaphor is so often attached to politicians, planners and other essentially short-sighted professionals. However, records kept over a period of 80 years at the ostrich farm of Oudsthoorn in South Africa, where 200,000 of the birds were reared in captivity for their feathers, showed not a single example of an ostrich manifesting this behaviour.

It was noted that they occasionally rested the head on the ground as though to rest the neck muscles. Other reports said that the ostrich could pick up ground-transmitted sounds by listening intently near the surface, and that they would forage around in the undergrowth. It also occurs to me that, since ostriches need small stones in the crop to help macerate their food (and this is typical of bird digestion) they would be seen from time to time feeling about in the topsoil for small pebbles and fragments of rock. One or more of these possible explanations may account for the legend, but it does not otherwise have a factual foundation.

WHY DO MICE LIKE CHEESE?

They don't, particularly. Mice will eat most organic materials that are not actually toxic (and some that are). The reason why cheese is always used on the spike of a mouse-trap is so obvious that I will not reveal the answer at once.

The reason why domestic animals eat specific foods has some connection with the evolutionary origins, of course; a cat will try to hunt birds through inborn instincts that are connected with the need to support itself in the wild, and many of these codes of behavioural criteria are inbuilt. But there are people who will tell you that their dog drinks nothing but tea, or their cat eats little but breakfast cereal.

The apparent faddishness has been taught to the animal by the owner. At some stage in early development, the animal has been presented with the now-liked foodstuff and the learning processes have in this way been instructed that a given substance is the right thing to eat. This is, of course, supported by a positive and encouraging attitude on the part of the owner, who is almost training the pet to take what is offered as a gesture of compliance and subservience. In this manner, specific likes and dislikes are taught to the animal, and its individual code of criteria (to use the model I apply to such learning discrimination) selects the specified food from then on.

The arbitrariness of the process is best illustrated by the 'standard diet' for cats. Most people immediately say that a cat's favourite food is fish, and its favourite drink is milk. A moment's reflection would make anyone realise that cats, which have a dislike of swimming, would not be likely to encounter fish in the wild. And how exactly they are meant to obtain milk from a cow under natural conditions is quite beyond my imagination! Here too we are witnessing the result of teaching, by an owner, at the critical phase of development when food preferences are being learnt.

So to return to the mouse and the cheese. It is obvious that there is no cheese in nature, so there will be no inborn preferences for that substance incorporated at genetic level into the mouse palate. Cheese is naturally an attractive food when mice find it in the larder, but then so are many other man-made items, such as bread, bacon or wheat.

The reason cheese has always been used on mouse-traps is because it is the only substance which has the right consistency to stay in position, and to actuate the release mechanism when the mouse tries to tug it free. Bread would be too crumbly, and could easily be removed. Bacon would not grip the spike, as it has no rubbery quality, and would be too easy to lift off. And quite how you would attach pearl barley, flour or desiccated coconut is beyond our capacity for initiative. Cheese is not chosen because mice have a particular preference for its flavour, then, but because everything else would not fit on, or would fall off if it did.

A mouse that survived being caught in a cheese-baited trap was once used as a miraculous portent for Austrian forces who, in 1796, were being beaten back by Napoleonic forces. The Austrians, at a loss to know which way to move, dipped the mouse's paws in ink and set it free on a plan of the campaign, in the belief that its divine capacity to survive would reveal the best route for attack or escape. In the event, the creature ran round in a confused path and left a shapeless set of tracings on the map. Prophetically enough, this is more or less what happened in the ensuing days,

and the Napoleonic forces soon won the battle. At the time the Austrian generals were disappointed that the mouse had failed to show them what to do; but in one way it was exactly right all the time.

WHY ARE BLACK HOLES SO MYSTERIOUS?

The essential reason why black holes are a mystery is because we cannot be positive they exist, and we could not see them even if they did.

In some 5,000 million years, our own star the sun will begin to cool to such an extent that it will no longer be able to support its own size, and the remaining matter will gradually expand (enveloping earth) before imploding. It is predicted that it will collapse inwards, like an explosion in reverse. It will then enter the stage in which it will be a highly dense star known as a cool dwarf. Larger stars than our sun will collapse into a still smaller volume, because of the greater gravitational attraction that the increased mass will exhibit. The interaction of the atomic particles thrust together in this catastrophic event will trigger a subsequent explosion outwards, and the result would be visible over thousands of light-years as a super-nova.

But if we carry this model to extremes, we can calculate that this exploding outwards could not occur if the initial star was massive enough. The collapse of one of the largest stars in the universe would bring together such a concentration of mass that the gravitational pull each particle would exert on every other particle would be too great for anything to escape. Not even the most energetic fragment of mass would be able to leave the gravitational field; indeed it is postulated that even light rays themselves would be drawn back into the centre. For this reason, no outward sign would betray the existence of this super-dense body to a watching astronomer. And it is this condensed form of matter that is dubbed a black hole.

So dense would the matter become that a block the size of a matchbox would weigh 10,000 million tons. A large star ten times bigger than our own sun would, under these forces, shrink to a sphere only 40 miles in diameter. Around the black hole would be found a zone where it was just possible for energy to escape from the gravitational centre. This critical zone is called an event horizon and it is associated with the many predicted oddities of behaviour that black holes would embody.

Because the black hole is forever turning inwards on itself, whilst the rest of the universe is doing the opposite and continuously expanding, it is claimed that the black hole is like the beginning of the galaxy, only in reverse. From this it has been argued that time would appear to run backwards in the area of a black hole (i.e., inside the event horizon). A spacemen entering this area might appear to remain on the surface of the event horizon for millions of years afterwards, his image trapped on that critical boundary. If he was watching a clock face as he approached this zone, he would – it is claimed – notice that the passing of time seemed to accelerate as he neared the event horizon, and at the edge it would be impossible to see the hands at all. From then on in, however, he would see that the time was now moving backwards, and the clock would spin backwards towards the beginning of time.

The gravitational attraction would drag him rapidly into the centre of the black hole. He would of course be instantly destroyed by the forces acting upon his spacecraft, but if you postulate that he landed on the surface of a black hole he would still be pulled through it and into the centre.

The pull of gravity would be so great that he would pass from the surface to the centre of a black hole twice as heavy as our sun in 20 millionths of a second. Even if the entire black hole was gigantic and, say, a million times larger than our sun the journey would take less than 10 seconds. Because of the nature of the black hole, the space traveller could send out no radio message nor any signal, and there would be no means in which – even if somehow he could contrive to stay in one piece – he would ever be able to escape from its gravitational field.

In summary, that is how the theory of black holes postulates they would behave. It is all theory, and some of it is contentious, but it has the makings of marvellously inventive stories about space, the future, time travel, and science fiction. The modern film industry needs black holes almost as much as the astronomers, all of whom are looking for funds harder than ever. One investigator has recently proposed that there is actually a small black hole inside the sun, helping it produce its characteristic pattern of radiation, and several have put forward the idea that there might be a black hole in our own galaxy.

Evidence in favour of the existence of black holes is tenuous, but it does exist. One example is the behaviour of the star Epsilon Aurigae, a giant star that is calculated to have a brightness 60,000 times that of the sun. It has been found that every 27 years it has a regular cycle of change in brightness, for it fades at those times to half its usual brilliance. It is known that Epsilon Aurigae is a binary system, for it has an invisible neighbour. The two of them orbit round a common centre of gravity, and when the partner gets in between the star and the observers on earth, it seems to become dimmer.

What the astronomers did was to demonstrate that the unseen partner emits no radio waves, nor does it send out any light in the way that other stars do. The answer that has been put forward for this apparent mystery is that the partner is a black hole, and so cannot emit any wavelengths that would betray its presence to us.

There is another example in the constellation Cygnus, which is catalogued as HDE 226868. This star is also part of a binary system, and it seems to emit a stream of luminous gas clouds which extend for a considerable distance, and then disappear. The explanation put forward is that it too has a black hole as neighbour, and the region where the gas clouds seem to vanish is the line of the event horizon, and from there on they can never be observed.

Black holes are intriguing to the philosopher since – by definition – we cannot observe them directly. The hypothesis does seem to account for the fate of giant stars, and yet only inference can provide evidence in favour of the idea. Not only can we not be sure at present that we have found a black hole, but we never can be sure. For the astronomer, it means he is on a never-ending quest.

WHAT IS A QUASAR?

The term was neatly devised to categorise a group of stellar bodies that do not behave like typical stars. They are more like quasi-stars, and quasars is as good a name as any for them.

The first quasar to be identified was discovered by a Dutch astronomer, Maarten Schmidt, working with a radio telescope at Mount Palomar in California. He was searching for the source of a powerful radio emission in the Milky Way, but what he found did not appear to be one of the Milky Way stars. Its spectrum showed it was moving away from us at a velocity of many thousands of miles per second, and it was calculated to be already 1,000 million light years (1 light year = 5,865,696,000,000 miles) away from us. The object was much brighter than a conventional star, and it started a search for others like it. This, the first quasar, was catalogued as Quasar 3C273, and the year of its discovery was 1963. Since then around 300 quasars have been discovered, the outermost travelling at speeds that approach that of light. Their energy output is prodigious. One quasar emits 200 times more light than all the stars in our own Milky Way galaxy combined. But nobody really knows *what* they are. Hence the 'quasi'.

HOW WERE PULSARS DISCOVERED?

The first pulsar was located by an astronomy graduate student named Jocelyn Bell who was working with a Cambridge University radio astronomy team in 1967. Professor Anthony Hewish, who headed the research team, described the radio signals as sounding like a regular, broadcast time-signal, and at first many people were tempted to conclude that they were listening in to a synthesised signal, transmitted across space from some other civilisation on the lookout for intelligent life elsewhere in their universe.

The signals were monitored over a period of months, and it was realised that they did not come from any intelligently-constructed transmitter, but were the result of a natural periodic radiation. The signals were received as pulses of predictable regularity, a little over one and one-third seconds between each pulse (actually 1.33730113 seconds).

Since then, several other pulsars – meaning 'pulsating stars' in scientific shorthand – have been found. In 1969 came the discovery of the first example that could be seen by optical telescopes, as well as detected by radio. Astronomers working at the Steward Observatory in Arizona discovered this example in the Crab Nebula, over 5,000 light years distance, and showed that its light emission varied in step with its pulsed radio transmissions.

Two theories were put forward to account for the regular transmissions. One of these was that the pulsars were spinning round, radiating energy in one main plane like a lighthouse with a revolving mirror. When the earth happened to come into the line of 'sight' of the pulsar, the argument went, we would pick up the beam. However, some of the transmissions have a frequency as high as thirty pulses per second, which seems to make rotation unlikely unless a complex pattern of radio beams is being emitted by the same source.

The other school of thought held that pulsars are in a peculiar state of activity, expanding and contracting as it were, and sending out a flickering series of energy pulses in tune with this periodicity. Here too there are problems, for the signals are clear-cut and not muffled, and if they were being produced by a star in the ordinary way then the signals from the far side would take fractionally longer to reach us than those emitted from the side of the star nearest to the observers on earth. The clarity of the signals suggests that we have a very small transmitter. How it would generate its pulsed signals, which at first sounded like a message from another world, remains to be decided.

WHERE IS THE EARTH IN SPACE?

In our orbit roughly 93 million miles from the sun we take $365\frac{1}{4}$ days to complete one cycle. The sun itself is a star 864,000 miles in diameter, its surface heated to 5,000°C by nuclear reactions. Our solar system – i.e. the earth and the other eight planets circling the sun – is about 5,000 million years old and mid-way through its life-span.

The sun is no more than an average star, and it lies in an outer limb of a spiral galaxy consisting of roughly 100,000 million stars. The galaxy itself is 100,000 light-years wide, and the solar system is found one-third of the way across. The galaxy is composed of a flattened disc of stars, shaped like two saucers placed together with edges touching. If you look at the night sky, it is easy to make out the Milky Way as a broad stripe of a misty white running across the sky from horizon to horizon. What you are looking at is the view through the plane of the galaxy – returning to the saucers analogy, you are inside, looking out towards the rims.

There are other galaxies like ours, the closest being NGC 891, which is about 20 million light years away. The most powerful telescopes can demonstrate the presence of a thousand million galaxies, and if they all contain 10,000 million stars, there are 10^{20} stars within our sight, or 100,000,000,000,000,000,000 to write it out in full. Put like this, the odds against there being other life somewhere would seem to be unrealistic. However, even if there are other civilisations in other galaxies it is difficult to see how we might make contact with them. Even within our own small corner of the universe it would take an unthinkable length of time to make the journey. The nearest star to the sun is Proxima Centauri, and that is 4.3 light-years away.

The galaxy itself is now roughly 10,000 million years old. In 5,000 million years' time the sun will undergo a dramatic change in size, bursting outwards to swallow up the orbits of all the planets, followed by shrinking away to become a dwarf star a few thousand million years after that. These are imponderable times, of course, yet they acquire an assimilable flavour by being represented in figures. The story is told of the drinker in a Manhattan bar who overheard a conversation about the demise of our world. The figures caught his attention, and he banged down his glass with a fearful look in his eyes.

'Pardon me,' he said to the astronomy enthusiast who had spoken, 'What did I think I heard you say?'

The man looked round and said, crisply: 'I was just telling my friends here that in a billion years all this will have vanished. The whole lot. Gone.'

A look of relief spread over the questioner's face. 'A billion years,' he said, 'thank God. For a moment I was real worried. I thought you said a *million*!'

DO ANIMALS REALLY USE THE STARS TO NAVIGATE?

Yes, they seem to. The trick answer to this question is to say that almost all navigating creatures know which way they are heading because of the position of one star. But that star is the sun, and it does not really answer the question in the way the questioner had in mind.

The first observation that drew attention to the likelihood that migration could be guided by stars was the fact that many birds migrate at night, and seem to do so with considerable accuracy. The influence of star patterns is relatively easy to investigate, since birds can be put into study cages and then taken to a planetarium where the exact view that the bird would have of the stars at night can be obtained at will. A German ornithologist named Franz Sauer first proved this by studying warblers. He found that if he rotated the position of the star images on the dome of the planetarium by 45 degrees, then the birds adjusted their own bearings by the same amount.

Surprisingly, even birds hatched in an incubator could make the same distinction, although they had never seen the real sky throughout their lives. But the response was not identical in all species. An American investigator, S T Emlen, studied indigo buntings that had been hand-reared, and found that they showed no orientational response like that exhibited by Sauer's warblers. But during his experiments he did leave the birds exposed to an experimental sky in which the normal rotational pole – near the pole star – was altered so that it rotated around the constellation Orion instead. Later, when they were exposed to an autumnal sky map, they showed a preference to orientate themselves in relation to their experiences under the artificial sky. Here was an example of their learning a code of criteria that satisfied an inborn curiosity, rather than (as was the case with the warblers) inheriting the full set of information on a genetic level.

Birds are not the only creatures that seem to use the stars as direction-finders. Moths migrate by night, and the possibility that they might be homing on a compass direction detected through the earth's magnetic field was disproved by painting over their eyes and releasing them under test conditions. If they were blinded, they lost their sense of direction at night. They might have been guided by signals on the wind, or by wind direction, yet this possibility was disproved by confining them in closed, dark rooms. Once again, their direction-finding capability was eliminated.

Even when the top two-thirds of the moths' eyes were painted over (thereby cutting out their view of the night sky) they were unable to navigate at night. But yellow underwing moths show an ability to navigate perfectly well on a moonlit night, and even on a night when there is no moon but when the sky is clear. Under cloudy conditions, however, they lack the ability to find their way. The most likely conclusion for these creatures too is that they understand star-maps well enough to know which way they are flying.

HOW FAR CAN BIRDS MIGRATE?

Most of us tend to think of traditional notions of bird migration, such as the seasonal disappearance of swifts and swallows in the autumn and their reappearance next spring, or the regular sound of migratory birds like the cuckoo that remind us how efficient the migratory timetable really is.

But the birds that undertake the greatest feats of navigation are species most of us never see, and they show navigational abilities, and sheer stamina, that belie their diminutive size.

The Arctic warbler (a relative of our own willow warbler) weighs about ten grams, less than one-third of an ounce. It breeds nowhere further south than Scandinavia and northern Russia and it is common in northern Norway. This bird finds its feeding quarters in the winter in south-east Asia, and so it has to undertake a journey of thousands of kilometers twice every year to survive in its two environments.

The wheatear sets an even more prodigious record. This little bird has its winter quarters in Africa, and it has followed the retreating line of the ice age in colonising northern Europe for thousands of years. This has meant that the distance between its summer and winter habitats has steadily increased. It has spread across Greenland to Canada, and – on the other side of the world – across Russia to the east of Siberia. It now seems that the birds that fly east in the summer almost meet up with those that fly west: the two groups in Alaska and Siberia are now known to be within 500 miles of each other, almost circling that part of the globe.

The record-holder, so far as distance goes, is the Arctic tern. This delicate, tapering-winged seabird is found in the Arctic, indeed within a few hundred kilometres of the North Pole. Yet it spends the spring in a 12,000 mile (20,000 km) migratory flight down to the Antarctic pack ice. By June or July it has turned, ready for its homeward flight, and by the winter it is back again in the northern polar regions. It spends two-thirds of each year on the wing, diving into the ocean to catch fish for food as it does so. Its distribution, ranging from near the North Pole to the south Atlantic and Pacific oceans, almost spans the globe. No other species migrates so far.

The European swallow spends the summer season breeding and feeding in northern latitudes and the winter in Africa. There are several groups of swallow related to the house-swallow: the mid-Asian swallow, which winters in India, is one separate group, and there are swallows in the far east of Asia that winter in south-east Asia and sometimes reach into Australia. The American

swallow is a different race of the same species as the European swallow, and it ranges across the United States and Canada, whilst the cliff swallow nests as far north as Alaska. In winter they fly south to warmer climates, the swallow reaching as far as Argentina and Chile.

In America, the northerly migration starts in February. The birds are in evidence in California by late March and the cliff swallow is back in Alaska by early May. The birds fly to the East Coast states as well, and those that make this journey take somewhat longer as they have to make up extra mileage by skirting the gulf of Mexico.

Swifts and swallows from Africa have reached Spain by early March. By the middle of the month they are into Italy and the south of France, and by mid-April they have returned to the southern half of Europe and reach up into northern France and the south-western areas of Britain. In early May they are back in Scotland, Denmark and the Soviet Union about as far north as Moscow. The house martin follows a little later, probably because it catches insects on the wing – like its cousins – but at a higher altitude than they do. Insects do not fly high until the air is warmer, and

for that reason martins seem to time their arrival for later than the swallow and the swift. Some of these birds have been shown to reach speeds of 150 mph and they can cover 650 miles in a day at slower, migration velocities when, in July and August, they begin the flight south, to repeat the journey next spring.

The cuckoo makes a similar flight, and it is familiar in Europe between April and August. Its earliest arrival in Britain was an observation on 2 March 1972, when it was seen as well as heard in Wantage, Berkshire. The latest sightings on record are in December – 16 December 1912 and as late as 26 December, in Torquay, from the late 1890s. There is an old rhyme that describes the cuckoo's behaviour:

'In June I change my tune,
in July away I fly,
in August away I must.'

But these extreme records suggest that some cuckoos have a mind of their own

The young birds, which as is well known are raised as 'parasites' in the nests of pipits, wagtails or other birds, in any event leave later than the parents and are often seen in September. It is interesting to note that they have an inborn code of criteria that enable them to select the right way to fly when their own migration comes, though how this complex system could operate remains a mystery.

It should be noted that the American cuckoo is a member of a different family, and is not really a cuckoo at all. Incidentally, the discovery of the fact that the cuckoo lays its eggs in other birds' nests was first promulgated by an investigator whose name we think of in another connection altogether. He was the discoverer of vaccination against smallpox, Dr Edward Jenner.

WHICH OTHER SPECIES UNDERGO MIGRATION?

The phenomenon is widespread in the animal world, for many species undertake seasonal movements between breeding grounds for the summer season and warmer climates in winter. There has been a tendency for some specialists in the field to widen the definition of the term to encompass any kind of movement from one place to another, under which umbrella term they also include normal patterns of territorial invasion (which is not migration) and even the shedding of seeds and spores by plants into the wind, an entirely unrelated phenomenon.

Butterflies are regular migrants. The red admiral and the painted lady are both familiar species that undertake long migrational journeys each year, covering many hundreds of miles in the process. The most famous of the migratory butterflies is the monarch, or milkweed. It is a large insect, one of the largest of the butterflies and it moves its wings so slowly and deliberately because of their surface area that it flaps its way through the air in a characteristic manner. The species is best known as a North American butterfly, though it sometimes takes advantage of the westerly jet stream in the upper atmosphere and in this way flies across the Atlantic, to be found in Britain and continental Europe.

Masses of these butterflies fly around 1500 miles each autumn from north America to their winter territory in Texas and Mexico. So precise is the

The red admiral and the painted lady . . .

moment of their departure that they all travel in huge clouds, and their passage is often marked by traditional celebrations along the route. Pacific Grove, California has even unofficially renamed itself 'Butterfly City, USA' to commemorate the fact.

A large number of the butterflies never reach the overwintering areas before the cold air masses of winter overtake them. They settle in huge swarms on the trunks of trees, flying when they are in warm air and settling back again to roost when the temperature falls too low to sustain them. The trees on which they settle are densely packed with the insects and look like Christmas garlands covered with the moving, orange wings.

Locusts are also well-known as migratory insects, as they take a long looping pathway circling clockwise across northern Africa. Normally they mate in areas where it has rained and where there is plenty of vegetation. The female then extends her abdomen to some four or five inches long, burrowing backwards with it into the sandy soil as she does so. With a pair of claspers at the end of the abdomen, she opens out a little chamber in which the eggs are deposited. As she withdraws her abdomen, she fills the tube that remains with a specially secreted foam. It is generally taught that this is to prevent the eggs from drying out, but that is not the reason for it. The eggs are deep in soil anyhow, and negligible moisture would be lost from a narrow tube. No, I find that the real purpose of the foam is to provide an easy way for the young locusts to emerge. They eat their way up through the foam in a matter of minutes, and its purpose is clearly to provide the young 'hoppers' with an easy way out.

Occasionally locusts breed in far larger numbers than normal. When this happens vast clouds of the insects can spread across large areas of land, eating almost everything in sight. Only the most resistant crops (such as cassava, p. 30) survive. Little is known about the triggers that initiate swarming, and the locust swarm remains one of the greatest threats to the farming communities of the underdeveloped countries.

In some cases it is easy to see what factors guide the migrant insects. Locusts, for example, seem to be led by the rainfall areas that change with the seasons. Migrating butterflies only fly in daylight when the sun is shining, since they need this external source of heat. In consequence, they have evolved sophisticated methods of relating their flight direction to the position of the sun over the horizon.

Even fish migrate over great distances. Salmon breed in the fresh-water streams of the Atlantic and Pacific seaboards and the young parr – the coloured juvenile form that grows as it swims downstream towards the sea – changes dramatically as it matures and begins to feed in the sea. At this stage, when it is known as a smolt, the salmon embarks on its life-time of hunting and feeding in the oceans, where it may live for three to twelve years in the case of Pacific salmon, one to four years for their Atlantic counterparts. At the end of this time, the adult salmon swim back towards their breeding grounds, usually back to the same stream (though clearly that is not an invariable rule, or new rivers would never have been colonised). During this journey the main organs of nutrition degenerate and are utilised by the adult salmon as its own internal food supply, so that by the time the fish are swimming up their chosen river to spawn, they have become little more than motivated containers of sperm and ova.

Pacific salmon invariably return to the sea after spawning and die shortly thereafter, whereas the Atlantic salmon usually survive for a little while, and a small percentage (probably less than ten percent) actually feed, regenerate, and return to breed again. One salmon was recorded after living for 13 years and breeding four times, but this is an exception.

Eels, as everyone knows, breed in the still waters of the Sargasso sea, a vast area in the Western Atlantic just north of the Tropic of Cancer. They develop as transparent larvae that drift in the northward currents and change into elvers as they enter coastal waters. They swim upstream, often crossing shallow water and – if necessary – leaving the water altogether to cross wet ground. They can slither through fields of grass and areas of undergrowth like snakes. Eventually they mature as they feed in the streams and rivers, or in enclosed lakes, and subsequently return down to the sea to commence their long swim across the oceans to the areas where they breed.

The extent of the invasion of fresh water by eels ranges along the east coast of America from the Guianas to Canada and on to southern Greenland, through the British and west European coastline and across the Mediterranean to the Black Sea. The shortest time taken by eels to reach their 'home' river is one year (for the east coast of the US) to four years (for the Baltic eels).

Whales migrate with the seasons, too. Humpbacks as an example regularly travel between the north Pacific and the Indian Ocean, which takes them around south-east Asia, a distance of 10,000 miles (16,000 km). These 40-ton adults can cover up to 100 miles in a day.

Land mammals also migrate, though they do not cover such prodigious distances. Caribou in northern Canada move a few hundred kilometres between winter and summer grounds, travelling in herds that may be as large as 100,000 individuals, but most examples of mammalian migration (such as the herds of gazelle, wildebeeste and zebra that regularly cross the Serengeti National Park in Africa) are leisurely phenomena. Fish, whales, flying insects and birds manage the task more efficiently because they rely on a fluid medium – air or water – to support their weight. Walking on land constitutes something of a disadvantage.

HOW DO MIGRATING CREATURES KNOW WHERE THEY ARE GOING?

There are many theories that attempt to explain this phenomenon, but all of them fail in the last resort. It is known that animals can orientate with regard to the sun, the moon, and even star maps (p. 62); and more recently tiny organs containing magnetic minerals have been located in some birds and mammals, and even in one group of motile bacteria. Internecine battles rage between supporters of one fashionable theory and another.

However I have no doubt that animals find their way in response to as many different inputs as we do ourselves. No-one would claim that man was motivated only by the direction of the sun, or the moon, the stars, or a magnetic sense. We find our way by a combination of factors, and turn to an alternative if one preferred mechanism is not available at the time. In fact if you were to think of the innumerable factors that you were actually aware of in your journey to work, or into town, this would reveal something of the great complexity of stimuli we rely on as inputs. Living organisms characteristically extract information from a combination of sources.

The problem remains, however, of how the animal uses these various lines of information to find out where it is. How does a young cuckoo, for example, know that it is supposed to head south and into Africa when it has never been there, and has no instructor to teach it the route? Knowing where it would receive its sensory input is no answer to that. And the amount that is known about how the animals use their various direction-finding abilities in a coordinated manner remains slight.

HOW MANY SPECIES OF LIFE EXIST ON EARTH?

Nobody knows, for there are large areas still not fully explored and there are innumerable new species of insect life, for example, still awaiting identification and naming, and imponderable numbers awaiting discovery.

Of all living animals known, only about 2% of species are vertebrates – that is, 'animals' in the popular sense of crocodiles, birds, guinea-pigs and so forth. The remainder are invertebrates, and include insects, worms, snails and shellfish. There are over a million species of insect known so far, and there are probably as many still awaiting identification, compared with a mere 3,200 mammals. A spoonful of water or soil rich in single-celled microbe life could easily contain more living individual organisms than the entire human population of the world put together. So the species that you think of as 'animals' are a tiny minority, and are hugely outnumbered by the so-called lower forms of life.

Two points of general interest I would like to add concern the incorrect way that species are often designated. First is the word 'species' itself. Many people speak of the term as though it were exclusively the plural, and use 'specie' as the singular. This is wrong. The word 'species' is a singular and plural noun, like sheep. It is no more correct to say that the singular of 'species' is 'specie' than to say that the singular of 'mice' is 'my'. Not only is that an error, but it is compounded by the fact that there is a word 'specie', which means coinage, as distinct from paper money.

The second problem that plagues one is the regular and widespread misprinting of species names. Sometimes they appear printed in roman letters, sometimes as capitals, sometimes in italics without an initial in capitals . . . the range is wide, and it extends through the breadth of scientific literature.

There is only one correct way to print species names, and it is this:

(1) the name is printed in italics;

(2) the initial letter of the first (generic) name is always a capital;

(3) if the genus is repeated in a passage of literature then it is normal to abbreviate the

which would be more correct. Another related group of bacteria are always abbreviated from *Streptococcus* to *Strep.*, which is also against the strict rule. The argument in support of this is that these genera of bacteria are often discussed

Tyrannosaurus rex

generic name to a single letter;

(4) when the name of the species is formed from the name of a scientist, explorer, discoverer or some other individual, then the second (specific) name may also have a capital.

This means that the human species is not Homo Sapiens or HOMO SAPIENS, but *Homo sapiens*. The term, in familiar usage, could be abbreviated to *H. sapiens*. The rock group *Tyrannosaurus rex* were exactly correct in abbreviating their title in later years to *T. rex*. If a Professor Smith were to discover a new species of *Tyrannosaurus* it might well be named *T. Smithii* or even *T. Smithensis* to commemorate the fact.

The only exceptions to the rule about abbreviations for the generic name are in odd cases where there is a tradition for a longer abbreviation to be used. There are only a few of these. The staphylococci are always abbreviated thus: *Staphylococcus aureus* becomes *Staph. aureus* rather than *S. aureus*,

together, and one would not want to mistake one *S.* for another.

Another rule-bending organism is a bacterium known to us all as *E. coli*. Strictly speaking, it should only appear in the abbreviated form when the full genus has already been spelt out once in full as *Escherichia*. But this proves to be such a mouthful to spell out, and it such a well-known organism (it inhabits the gut, and is the organism responsible for the high proportion of bacterial protein in faeces, p. 25) that people use *E. coli* instead. Not too many folk are quite sure how to pronounce the name in full, and even fewer are sure how to spell it.

You might also note that if you are talking about a particular species of a genus, but you are not sure which one it is, then you use the abbreviation 'sp' (for species, singular). On the other hand, when writing about the various different species of a genus, but without stipulating them, you would use 'spp', for species, plural. Thus you would write as follows:

'The human species *Homo sapiens* has a population of intestinal bacteria, including *E. coli* and *Lactobacillus* spp. There are more of these in a spoonful of shit than there are *H. sapiens* in the world.'

Indelicate, perhaps. But accurate. Note that it is wrong to speak of 'the *Lactobacillus*'. The genus and species names are names, nothing else, and should be regarded as such. Our dinosaur-discovering professor (p. 68) would not expect you to say that the species had been discovered by 'the Smith'. Similarly, the words Pete Shelley sang in a recent hit – 'be a Homo sapien too' – are wrong. It's always 'sapiens'.

WHY DO BIOLOGISTS USE LATIN NAMES?

For clarity and unambiguity. The problem with non-Latin names (to be precise, non-scientific names, for they are as often Greek in derivation, but I will follow popular usage here) is that they vary from place to place and from organism to organism. The bullrush, for instance, is not a bullrush at all but a reedmace. A green plover, a peewit and a lapwing are all the same species. A Christmas rose is not a rose at all, whilst a hybrid tea is a rose and not a blended beverage . . . and so on. Names also go out of fashion. Hybrid tea is at present due to be displaced by an up-to-date term, for example, and wandering jew plants are being renamed something (like spiderwort) with a less divisive sound.

In the mediaeval period plants and animals were often designated by Latin or Greek terms, but these tended to become lengthy and cumbersome and were sometimes more a short sentence than a long name. The introduction of a sensible form of classification, with two names – one for the genus, one for the species – has taken place since it was first proposed by the Swedish naturalist Linnaeus (1707–1778) in 1728. The names he proposed for many common plants and animals are still in use. In formal texts, though not in general works, the species of an organism is sometimes printed with an abbreviated version of the name of the person who so designated it. Thus Linnaeus's name is abbreviated to 'Linn.' or 'L.' and this addendum is widely found in reference works on botany and zoology.

The principal benefit is that there will be no doubt about which organism is under discussion. There is an added bonus in that the scientific name often describes the species in some detail. Thus *Lactobacillus acidophilus* means 'rod-shaped milk organism which likes acid conditions'. Our own species *Homo sapiens* means 'Mankind who is wise', and *Staphylococcus aureus* means 'little round organisms in groups like bunches of grapes of a golden colour' whilst *Tyrannosaurus rex* means, depending on your point of view, 'king of the tyrant lizards' (or possibly 'one of the most innovative groups in history').

CAN SITTING ON A COLD WALL GIVE YOU PILES?

In piles, or haemorrhoids, enlarged varicose-type veins form inside the rectum. They cause discomfort and may bleed. This is a dangerous and painful affliction and it is unfortunate that people find it faintly amusing. The cause of piles is an obstruction to the blood circulation in the affected area. Sitting on a cold wall would not cause piles. It might, though, relieve the discomfort.

WHAT IS BOTULISM? WHAT CAN I DO IF I SUSPECT A CAN OF FISH MIGHT BE CONTAMINATED BY IT?

Botulism is a form of food poisoning that is highly dangerous. It is non-infectious. The organism that causes botulism is one of the gangrene group and is called *Clostridium botulinum*; it produces probably the most poisonous toxin known to science. As little as a pound in weight of this toxin could theoretically kill every man, woman and child on the planet.

The bacterium itself is $\frac{1}{100}$ millimetre long, shaped like a straight sausage with rounded ends. The organisms grow singly or in pairs, end to end, and they can swim actively by means of fine, whip-like flagella that thrash in the surrounding water. There are at least six types, lettered A–F. Only types A, B and E are known to be associated with botulism in humans.

C. botulinum is an exception to one of the fundamental rules that govern life as the layman understands it, for this organism can exist without oxygen. Indeed it is killed – or at least inactivated – in the presence of even a trace of this (to you) life-giving gas. It prefers an environment that is about blood-heat or a few degrees cooler, though Type E can grow near freezing point. As the species grows it releases a toxin known as botulin. It is this which kills so effectively. Botulin prevents the transmission of nervous impulses, so that interference with breathing and total paralysis occur.

Because botulism and oxygen do not mix, the organism cannot exist in the healthy tissues of a mammal, such as man. Its ideal environment is in an oxygen-free substrate such as dead and decaying flesh. The bacteria can form resistant spores, microscopic hard-walled egg-shaped bodies that can withstand boiling. Then, they are able to germinate and resume their life-cycle when suitable oxygen-free conditions recur.

Botulism is usually associated with tinned foods, particularly tinned fish (often salmon). Unless the heating process is carried on during canning for long enough, the spores can survive and may germinate inside the sealed can. Alternatively, it is possible for a leaking seal to allow a trace of contaminated water to gain access to the can. This could introduce viable spores which were later able to germinate. Quality control makes such events extremely rare, but there are occasional reports which call for the return of a batch of cans. As canned goods are often prepared on contract at a central plant acting for several distributors who operate with their own brand-names, the same batch may not be recognised from the name on the tin, but from the code number embossed on the lid.

The question of what to do about a possibly contaminated can worries many people whenever there is a botulism scare. If you have a can of, say, salmon and a warning has been issued, then it is instinctive to be concerned about cans that do not even bear the suspected code. They cannot be returned to the store, as these are not the ones that have been recalled. But some people remain worried about using them. However, there is a reliable way of making canned foods safe.

I have not tried this myself, but the bacteriology authorities who deal with this organism have found over the years that botulin is destroyed by boiling. Indeed, the strange fact is that the chemical these bacteria produce is more easily destroyed by heat than are the (living) spores from which the organism emerges. To kill botulism spores, you need a temperature of 120°C. To inactive botulin, however, a temperature of 90°C for 40 minutes will suffice.

To put your mind at rest, therefore, you could decide to use up canned fish about which you are otherwise going to be worried in a recipe that calls for cooking of this kind. Alternatively, standing the can in a covered container of boiling water for an hour would have the same effect. Both of these treatment schemes would enable you to feel sure that any traces of botulin had been destroyed.

Of course, if a batch of contaminated cans is announced by the news media then it is in your own interests to return any cans with the right batch number. They can be checked by the company in due course, which will provide useful data on the nature and extent of the problem, and if cans are recalled in this way there is no financial penalty on the consumer – you receive your money back.

CAN FOOD POISONING BE AVOIDED, AND IF SO, HOW?

Food poisoning labours under a variety of titles, ranging from the elegant 'gastro-enteritis' through the dramatic 'dysentery' to the 'squits', 'runs', or 'Montezuma's Revenge' (named after the Aztec emperor who greeted Cortez when he invaded Mexico in 1519, and who was betrayed).

There are two main categories of food poisoning, one caused by disease-causing bacteria, the other caused by toxins produced by bacterial growth. The main bacterium responsible for the first category of food poisoning is *Salmonella* spp., and it is *Staphylococcus* growing in foodstuffs such as whipped cream that produces the toxins that comprise the second category.

The genus *Salmonella* includes several species, including *S. typhi*, the causative organism of typhoid, *S. paratyphi* which causes paratyphoid, and the usual source of the problem with food-poisoning, *S. typhimurium*. There are many different types of *Salmonella* which are known to cause food poisoning, and hundreds are routinely identified. In Britain alone there may well be 100,000 cases of food poisoning per year, many of them never reported to doctors.

Most people will know the symptoms of the disease. They include pains in the abdomen, vomiting and diarrhoea, and the passing of odd-coloured faeces which may seem almost green. The bacteria cause an infection in the digestive tract, and as they grow they release a toxin which is responsible for the distressing symptoms. Unlike the toxin of *Clostridium botulinum*, (p. 69), this substance is able to withstand 100°C for 30 minutes.

The essential factor in the development of this form of food-poisoning organism is that the bacteria need a warm environment in which to grow. They often occur in meat, but are widespread in other foods too, in numbers that are too small to cause any problems for the consumer. But if the food is left around to stand at a temperature which is amenable to the growth of the bacteria, then a large population can develop in a few hours, and a vast population overnight.

It is therefore important to understand this need of the organisms for warmth. The first and most direct method of ensuring that food poisoning is avoided in the home is to use hygienic food from a reputable and hygienic source. This is most important. Sliced, cooked meats that have been standing around in a fly-ridden store are not suitable for human consumption, and not really suitable for animal consumption either if you are any kind of pet-lover.

The second point is to see that the food is kept cool. In the refrigerator it is difficult for most bacteria of this kind to grow.

When cooking, you must ensure that the food is really cooked properly. This does not apply to a rare steak, for the outer regions here are well cooked by the heat of the fire whilst the interior remains partly raw. However, this poses no problem in practice, since any bacteria will be most likely to reside on the outside of the steak, and the centre will be sterile anyway. But larger food items, like a turkey, are often undercooked so that the centre can become warm – not hot enough to kill the organisms, but quite warm enough to encourage them to reproduce at speed.

Once the food has been cooked, it is helpful to reduce the temperature as rapidly as you sensibly can. It need not be refrigerated at once, indeed it should not be put into a refrigerator until it is quite cold or it may warm adjacent foods as well as overloading the cooling apparatus. As soon as it is cold, then, it should go into the fridge.

Finally, and most important, the food should not be allowed to stand in a warm place for any length of time. If it is to be eaten cold, then it should be served from the fridge and left for not more than an hour to acclimatise to the air. It is not gastronomically satisfying to serve food which still has the chill in it, unless you are thinking of fresh vegetables; sliced meats should be allowed to return to room temperature and an hour is perfect timing for that. It does not allow any organisms time to reproduce enough to cause problems.

The public are always warned never to re-heat food. The reason for this is that food which is warmed and left to stand can be incubating bacteria. As you can see, with some understanding of the way the organisms grow in mind, it is entirely possible to reheat food. But one of two methods should be adopted. Either the food should be thoroughly and rapidly heated through (perhaps by heating sliced meat in a gravy sauce at boiling point in the oven), or it should be warmed and then served at once. So long as you can keep less than an hour between taking the food from the fridge and eating it, whether it is re-heated or not, it is almost impossible to engineer a way in which food poisoning organisms could develop in the time.

The second type of food poisoning is caused by the toxins of *Staphylococcus* spp. The bacteria themselves do not usually survive passing through the acid of the gastric secretions (p. 24) and so they do not cause an actual infection of the intestines. Instead, they grow in the foodstuffs and produce toxins which are themselves absorbed as the food is digested. These toxins are the most heat-resistant of all those we have considered: a temperature of $125\,^{\circ}\mathrm{C}$ for one hour is necessary before they will be inactivated.

These organisms will grow in many foodstuffs but they favour a medium where there are some carbohydrates available. Trifles are a common source of trouble, for they will be made with cream, sponge and all manner of nutriments that are much to the liking of the bacteria. The organisms themselves are common inhabitants of the nose, the throat, and the skin, and so there is ample opportunity for them to gain access to the foodstuffs under preparation.

The answer here is to try to keep to an absolute minimum the length of time between preparation and consumption. If that is impractical, as it often

is, then the made-up foods should be put immediately into the refrigerator where the organisms will not grow. Once the food is well cooled it can be stored for a while, and should then be left for less than an hour before being eaten.

The point about the two categories of food poisoning is that the one is essentially an infection whereas the other is poisoning caused by a toxin. Clearly, therefore, food poisoning from *Salmonella* would be expected to take time to develop. The organisms need to reproduce themselves and establish a community, and that can take perhaps 15 hours. The toxins produced by *Staphylococcus*, on the other hand, are already present in the food and therefore you would expect that form to develop much more rapidly. The toxins need time to be absorbed by the bloodstream, but that can take place within the hour.

This picture is complicated by the occasional cases of *Salmonella* infections where the symptoms also develop within a very short time. These are due to the organisms' growing to considerable numbers within the food, so that they have produced a reservoir of toxin that is available to act at the moment of absorption. But in cases like this, the food must have been left standing in a warm situation for some time.

Naturally we have here an outline of just a few kinds of food poisoning. There are many organisms that can grow in food (hardly surprising, since we have made the food to be nutritious and it would be strange if we were the only species that found it so), and they cause a range of conditions it is sometimes hard to identify accurately.

But with the principles outlined above it should be possible to avoid almost all of the types. The cardinal rule is never to leave food for long periods in a warm place. Without warmth, the organisms that cause food poisoning will not rapidly grow. And if they cannot build up a sizeable population, then there is no risk of contracting any of the unpleasant conditions that we know collectively as food poisoning. The conditions have been steadily on the increase in many western countries, almost approaching an epidemic in the summer months. Yet we have seen just how easy it would be to control it.

A VIRUS: WHAT IS IT?

This term 'virus' probably occupies a unique place in human affairs. It is one everyone uses in the sense: 'If you ask me, you must have gone down with a virus!' yet the overwhelming majority of people have not the least idea what they mean by the term. Newspapers, and even highbrow magazines, regularly use the term to describe something that is not a virus. Having said this, it must be admitted that explaining the nature of viruses to anyone who has not spent a prolonged time studying microscopic organisms is difficult. Given that we all suffer from viruses, and that they have done much to mould human history and to alter the tide of mankind's political affairs through the ebb and flow of virus epidemics, they ought to be one of the basic things we learn about at school.

A virus is not a living entity. Living organisms can respire, and undergo complex energy-giving metabolic reactions that we recognise as nutrition. They reproduce by cell division (this applies to man, although we do not ordinarily think of it like that) and they feed on their surroundings. They grow, too.

Viruses do not act like this. They are very much smaller than even the smallest microbe. They do not feed in the sense that anyone would recognise as feeding, for they only possess the simplest genetic material and the few types that produce enzymes only do so to help them invade the host cell. This is the point about viruses: they all need a host cell. They do not have any of the normal cell machinery themselves, so they commandeer the equipment of a normal living cell, and instruct it genetically to start producing more viruses like the infecting agent. The dividing of microbe cells in two is a form of reproduction that passes on through the animal and plant kingdoms to the highest level; but this is a totally different process from the way viruses multiply.

The unique property of making a cell produce more of the virus, rather than the virus undertaking to reproduce itself, is known as replication. A virus is rather like a wandering parcel of genetic material, a 'chromosome hobo'. Because of the simple nature of viruses, some of them can be purified and crystallised rather like any simple chemical compound. This makes them difficult to study in some ways, for a virus must have a living cell in which to replicate, so it cannot be grown pure on a bed of nutrient jelly, as can living microbes. Furthermore, the fact that viruses replicate inside living cells makes them difficult to control. Drugs that would prevent replication also interfere with the rest of the cell's productive machinery, so materials that cure virus diseases tend also to kill the patient. That is why routine drugs are not available for virus infections.

In biology one is often asked: 'Why cannot science cure the common cold?' The answer is that we cannot actually cure smallpox, polio, measles, mumps, colds, influenza or any other virus disease. Many of them can be *prevented* by the use of vaccines that pre-charge the blood stream with antibodies. In this manner the immune system (governed by the white cells in the blood-stream, p. 47) can inactivate a virus as soon as it invades the body. But once a virus is there, and has started a disease process, there is no effective cure and it is largely up to the individual victim whether they succumb to the effects or not.

In a few instances, physicians have utilised concentrations of antibodies as a serum injection and this form of intensive treatment has been used in patients with such dangerous diseases as lassa fever. But this is a rare application, and even it does not amount to an effective treatment. Antibiotics, which are often prescribed for the victims of virus infections, are of no value in fighting the disease itself. They are administered to keep at bay any opportunistic bacteria who might use the body's lowered resistance as a cue to set off a secondary infection. This can happen in influenza, for example, when a bedridden patient who is beginning to recover finds that bacteria in the respiratory system are threatening pneumonia.

One of the greatest mysteries about viruses is how they produce the characteristic signs of their own specific diseases. Here I do not have in mind poliomyelitis, for example, for that is a virus that infects the nerves and progressively inactivates the central nervous system, causing a rising paralysis in its worst forms. The disablement of the nervous system is a clear consequence of the virus 'take-over' of the nerve cells, and the symptoms that develop are an obvious result of that effect.

The question-mark hangs over diseases such as smallpox, measles and chicken-pox (more officially known respectively as variola major, morbilli – I wager you did not know that before! – and varicella) in which the characteristic spots appear. The spots are recognisable, their distribution, size and shape are largely characteristic, and I find it perplexing to try and relate this occurrence to the viruses' action on the body. How they are transmitted to the skin cells and produce their own variety of eruptions would make a fascinating revelation, for it seems impossible to understand as things stand at present.

Is it possible that you are right to say, 'It is just a virus', when some unexpected mild infection supervenes? It may be that you are. There are many groups of viruses that can produce latent infections, for instance; one example is the adenovirus group first isolated from the adenoids. There are viruses that cause sores for a time, but normally remain hidden and silent in their effects – and some of these may be involved in apparently unrelated diseases. The herpes virus is an example of this category. The parvoviruses, also known as picodnaviruses, are not yet understood fully, but cause diseases in animals and may do so in mankind, whilst the papovavirus group (which have not been shown to have this effect in humans) seem to cause tumours in animals and some may cause warts. Reoviruses were so named as Respiratory Enteric Orphan viruses, because they seem to have nowhere to go. They are found in the respiratory tract and the intestines (p. 25) but are not known to cause any specific disease.

The picornaviruses include the common cold, polio and foot-and-mouth disease of cattle, a condition that can also infect humans. Yellow fever is one of the range of insect-transmitted diseases that comprise the arbovirus group – standing for Arthropod-borne viruses – whilst rabies is the most prominent of the rhabdoviruses. Other forms are the myxovirus, paramyxovirus and the well-known pox virus groups.

Between them they can cause a range of illnesses, some severe, most not, and an unknown number very mild indeed. Vaccines for some

"The vet's due at the Parthenon this afternoon"

previously intractable diseases are now available – a safe and reliable vaccine for rabies is one recent arrival on the scene – whilst others, like a vaccine against the common cold, seem almost as far away as ever as the viruses themselves are able to change their genetic makeup and so are not affected by antibodies produced for a slightly different strain.

Much remains to be discovered about the viruses. For example, human leukaemia (usually described as a cancer of the white blood cells, which are produced in excessive amounts, cf p. 47) is not known to be due to a virus, but all the forms of leukaemia that have been exhaustively studied in animals are known to be due to viruses. If a virus were to be implicated, then the possibility exists that a vaccine might be produced that would prevent human leukaemia. Viruses have been found in some forms of cancer, but this does not mean that the virus has caused the tumour. The viruses may be coincidentally present, they may be a result of the cancerous process, they could be produced in some way by cancerous cells. In this area of research we have to tread warily: the question must not just be: 'Do viruses cause cancer?' but should also be: 'Do cancers cause viruses?'

DO VIRUSES INFECT PLANTS AS WELL AS ANIMALS?

Some of the viruses do indeed infect plants, with results that range from being beautiful to economically disastrous. The most attractive example of a virus disease of plants is the break in colour of tulips that was so often painted by Rembrandt. This is not a genetic variant, but is due to a virus infection of the plant. A similar example is a virus (normally found in cabbages) which can be transferred by greenfly to wallflowers, normally of a plain colour. Many examples have occurred where a specific colour flower (a deep red, for example) has begun to produce mottled or streaked flowers that are yellow as well as red. This used to be thought of as a kind of 'genetic degeneration' in the seed line, but has since been shown to be due to a virus.

The tobacco mosaic virus has in previous decades caused economically disastrous damage to crops. One source was cigarettes, smoked by the growers. Viruses from the tobacco would be transferred on the fingers to the leaves of the growing crop, and an outbreak ensued. Similarly, the progressive degeneration of potato crops used to be thought of as the strain 'going stale'. There were objections to that, mainly the fact that in warmer latitudes (such as the south of England) the degeneration set in rapidly, whilst in colder climates (as in Scotland) the potatoes seemed to remain in good condition. As a consequence of this, Scottish seed potatoes were for years regarded as the safest source to purchase. The reason has turned out to be the virtual absence from Scotland of the greenfly that transmits the virus from an infected plant to a healthy one. Great lengths are gone to in modern nurseries to ensure that plants are virus-free.

Some viruses infect other forms of life, including bacteria, fungi and algae. Insects are particularly liable to virus diseases, and in some instances sprays of powder containing virus have been used to control insect pests that would otherwise decimate crops.

HOW WERE VIRUSES DISCOVERED?

After the upsurge in interest in the germ theory of disease, which began in the last quarter of the 19th century, it was accepted that diseases were caused by living microbes. But some clues began to appear that suggested that this might not be the case for all infections, and in 1892 a Soviet scientist showed that tobacco mosaic disease (q.v.) could be transmitted to a healthy tobacco plant by rubbing their leaves with juice from an infected plant. To prove that bacteria were not directly involved, he forced the juice through a filter made of unglazed porcelain, which contains tiny pores far too small to allow even the most diminutive bacterium through.

Oddly enough, in spite of the clear-cut evidence of his results, this research worker – named Iwanowski – persisted in the accepted belief that the disease was due to bacteria, and he tended to disregard the experiments with filtered tobacco juice, believing he had made a mistake. Six years later Beijerinck realised that an infectious agent that was too small to be a bacterium must be the cause. He labelled it a '*contagium vivum fluidum*' and, for shorthand, called the infectious agency a virus. He did not appreciate the modern significance of the term. The word means, in Latin, 'a poison' and it was in that sense the word was used. Most of the early workers used the fuller expression 'filterable virus' and it was not until the 1920s that the single term we use today was widely adopted.

By the end of the 19th century, foot-and-mouth disease had been shown to be due to a 'filterable virus' and so had myxomatosis in rabbits. In 1915 the first observations were made on cultures of bacteria that seemed to undergo dissolution. This was soon shown to be due to viruses that destroy the bacterial cells, and they were named bacteriophages, normally abbreviated to the simpler 'phage' and often denoted by the Greek letter φ. Thus, one of the phage viruses that infects the intestinal organism *E. coli* (discussed on pp. 25

and 68) is known as 'coliphage $\varphi X 174$'.

Because viruses are not living organisms they have not as a rule been given Latin names (p. 67) but are denoted by abbreviations and code numbers. However, one of the largest of the viruses, the causative agent of smallpox (a disease now extinct, at least outside a few virus laboratories) was given a Latin name along the lines of Linnaeus's binomial system. Few people used it, and it has now been forgotten. It was *Buistia pascheni* after Buist, an Edinburgh microbiologist, who first glimpsed it through an optical microscope in 1886. Most viruses could never have been seen by anything but an electron microscope, but in the late 1800s microscopy was at its height and optical microscopes were employed with far better effect than is the case today. Half a century after Buist's classical demonstrations, it was still being stated that 'no virus can be seen with the optical microscope'. Well, that may have been true of the 'advanced' 20th century, but back in the primitive old Victorian days there were some people who could tell their asymptote from their elbow . . .

CAN INTERFERON BE USED TO TREAT VIRUS DISEASES?

It ought to be possible, but so far it isn't. Interferon is produced as a reaction of the body's cells to an invading virus. It interferes with the replication process, or with the release of viruses from an infected cell, and was given its appropriate name by Isaacs and Lindemann, who discovered it in 1957. Aleck Isaacs went on to show that interferon did not have a toxic effect on cells at the dosages that were needed to prevent viral replication, it did not stimulate the production of antibodies against itself, and it seemed to have an effect against a considerable range of viruses.

It has since been shown, unfortunately, that there are innumerable interferons; some say there are as many interferons as strains of virus, and they are the optimists, for the pessimistic school believes there are as many interferons as there are

individual hosts. Interferon does not act on a virus when it is outside a cell, i.e. when it is actually being transmitted from one host cell to another. It has no effect on the way viruses attach themselves to host cells or penetrate into them, neither is it taken up by cells in significant amounts. If it is introduced into a cell, then it does not prevent the cell's death from the virus action, though it does seem to prevent the liberation of replicated virus from the dead or dying cell.

Isaacs himself, who did so much pioneering work in this field, knows nothing of these drawbacks. Driven hard by his work, and decried by many colleagues who did not understand his results, he became mentally unstable and died in 1962. The main thrust of research into interferon has more recently been cancer research, where the action seems to be against the division of malignant cells. However, this is an early stage in a difficult field of research, and in treating virus diseases interferon may yet be useful.

But it is ironic to think that when interferon was being hailed as the ultimate answer to cancer (something that is a wildly premature claim, to say the very least) the media spoke rapturously of this 'new wonder drug', little knowing that it was already twenty years old, and had failed to live up to its considerable promise first time round.

HOW BIG IS A VIRUS? A BACTERIUM? A LIVING CELL?

Surprising as it may seem, few microbiologists have 'total vision' for the relative size of the organisms with which they work. The society in which we live is appallingly ignorant about scientific matters, and biology in particular, on what you might call a 'general knowledge' front. I have been asked about the actual size of organisms far more often by non-scientists than by students of microbiology, and in talking to the latter group I am frequently surprised by the estimates they come up with.

In the following description it would be as well if you think of (or, even better, look at) a pair of round objects of which one is ten times the diameter of the other. A penny and a saucer, for example. Using these as a conceptual aid, we can now move gently down through the scale of life-size. By carrying out the same process in reverse, you will quickly be able to conceive of the relative sizes of the microscopic life-forms that surround us, and of some of their components. This is, believe me, a fascinating exercise and it is one that is rarely available in this form.

If we consider the saucer and the penny, the first step is to imagine the penny to be as big as the saucer. On this scale, the real penny is proportionately the size of a 2 mm foraminiferan shell. These are the rounded organisms out of which the massive chalk cliffs of the coastline are made. A little under half that size, 0.75 mm, is the size of a rounded diatom. These are plants, numerically amongst the most widespread of all the living things on this earth (p. 67), which grow a delicate internal skeleton of silica glass. We are now at the upper size scale of the world of microbes, and our shrinking journey can continue in earnest.

Imagine the diatom to be the size of the saucer. The penny lying on it is now 0.075 mm across. It would be easier here if I mentioned that the unit of measurement we use for these orders of size is the micrometre, written μm, which is one thousandth of a millimetre. On that scale, the diatom was 750 μm across, and we are now down to 75 μm. This is the size of a section across a bristle from a beard. A human egg-cell is twice this size. It is the bulkiest cell in the body, and, yes, it is easily visible with the naked eye, as a tiny white speck.

Now envisage the slice across the end of the beard-bristle the size of the saucer. The penny is now an object 7.5 μm in diameter. This is the size of an erythrocyte, or red cell (p. 47). I am naturally well aware that the textbooks say a red cell is 7.2 μm in diameter and not 7.5 μm, but they only refer to dead and dried cells. The live erythrocyte is indeed nearer 7.5 μm across on average.

Make the red cell the size of the saucer, and the penny is now the size of a small bacterium. If the bacterium is the saucer, then the penny is a chicken tumour virus. With this virus as big as the saucer, then the penny is just smaller than the tiny virus of

foot-and-mouth disease. A large molecule, like that of the chemical haemoglobin which is found in the erythrocytes, is three-quarters as big as that. And we are now in the realm of chemicals.

So there we are, a swoop through the nature of life. Each time going up by a factor of ten, we range from a haemoglobin molecule to an average virus, then to a bacterium, a red cell from the human blood-stream, the slice from a bristle, and a diatom. Double the size of that and we move to a chalk-forming microbe, then to the penny, on to the saucer, life-sized; ten times that diameter is the height of a man.

What an impressive complexity of organisation we here witness; and how fittingly into perspective our own numerically insignificant populations shrink.

WHAT DOES 'pH' MEAN?

There is a considerable fashion for what you could call 'pH-consciousness', and people seem interested in the pH of everything from their blood and urine to the soil around the roots of a potted plant and hair shampoo. But (as is often the case) without any idea of what pH actually means.

Chemicals tend to act in one of three ways: they are neutral (which applies to most materials), or they are acids, or they are alkalis. You tend to think of acids as being things that burn, but remember that there is hydrochloric acid in your stomach that is as strong as a bottle of the same acid in a laboratory (p. 24). And strong alkalis will burn just as much. Caustic soda (sodium hydroxide, NaOH) is a powerfully corrosive alkali and is a popular agent for cleaning blocked drains, or removing the most recalcitrant residues from a greasy oven. Just because it is an alkali, which is the converse of an acid, does not make it less likely to burn. So if you use caustic soda in the kitchen rubber gloves are an obvious precaution.

Defining an acid and an alkali is a matter that only concerns you if your work involves chemistry, and there is no need to feel that this is something that has to be understood in detail unless you work in that field. The majority of people who use a car with overdrive, for instance, have no understanding of how that works, exactly; they don't need to know, because that is a specialist topic of interest. The pH factor is likewise a specialist topic, but that doesn't make it a tremendously difficult matter to grasp.

The old idea on the definition of an acid and an alkali was to say that acids furnish hydrogen ions (H^+) whilst alkalis furnish hydroxyl ions (OH^-). The simple idea taught in schools is that an acid and an alkali (or an acid and a base, since alkalis are also known as bases) react to give a salt and water. Now you can see why. The acid provides the H^+, the alkali the OH^-, and together they make water: H_2O. In this case the reaction can be drawn out as follows:

$$HCl \quad + \quad NaOH \quad \rightarrow \quad NaCl \quad + \quad H_2O$$

Hydrochloric acid   sodium hydroxide   sodium chloride   water

and in this way the powerful, corrosive hydrochloric acid reacts with the dangerously caustic alkali to give that most neutral, everyday substance sodium chloride – the salt used in cooking and on your table – and water.

The problem with the definition as it stands is that it does not take account of solutions dissolved in solutes other than water, nor of the interesting exceptions to the rule. For instance, ammonia will react with water in the following way:

$$NH_3 \quad + \quad H_2O \quad \rightarrow \quad NH_4^+ \quad + \quad OH^-$$

ammonia   water   ammonia ion   hydroxyl ion

and in this case the water, H_2O, is acting as the 'acid'.

A more modern definition of an acid and an alkali is that an acid is a donor of protons, and an alkali is a proton acceptor. A proton is a hydrogen ion, H^+, so this definition retains some of its earlier counterpart. More recently still have come the so-called hard and soft acids and alkalis, discussed in terms of the free electrons in each substance. But this argument becomes increasingly abstruse as it becomes irrelevant for everyday purposes.

Like the difference between a dry or a sweet wine, a racing or a touring pedal-cycle, or jam and marmalade, you know one when you see one. And if you do not come across these shades of meaning in your daily lives, then there is no reason to feel apologetic if the complexities of the issue remain beyond your grasp.

But you might well want to know how to measure the relative degree of acidity or alkalinity of a substance. You might put a strong alkali at ten on a scale, for instance, with a strong acid at nought. On that basis, a neutral liquid (like water) would rate at five. This is a reasonable enough starting point, but there is a practical difficulty. If you have a strong alkali rating at, say, 10 on your scale, it is in practical terms not very different from a strong alkali rating 9 or 8. Anything from, say, 7 upwards would be a 'strong alkali'. Similarly, any acid rating 0, 1, 2 or 3 would be a 'strong acid'.

The problems arise because when you are around the neutral mark, rating 5 on this scale, even the slightest change in alkalinity or acidity can make a vast difference. For instance, if the body's tissue fluid ever fell below 5 by even the minutest amount, a condition known as acidosis would set in and you would have a sick person on your hands. On the other hand, if the tissue fluids are a small amount higher than 5, the victim would develop tetany and be thrown into muscular spasms. The body has to remain at something like 5.0001 on this scale for health to be maintained. Even at 4.9 or 5.1 you end up with a corpse.

What was therefore necessary was a means of measuring acidity and alkalinity that fitted this practical limitation. The scale had to be 'compressed' at the extreme ends, so that strong acids and alkalis were measured with less precision; and at the same time it had to be 'expanded' around the middle, so that the critically important slight differences could be spread out and measured with great accuracy. It was proposed that this should be done by taking the hydrogen ion concentration of a substance, and then converting that to the logarithm (to base 10) of its reciprocal. The fact that we are measuring the hydrogen ion concentration as a reciprocal means that when the hydrogen ion concentration is very high – as in a strong acid – the reciprocal is therefore very low. So a strong acid on this scale rates at near zero.

The converse applies to alkalis.

This mathematical trick is nothing more than an arbitrary subterfuge designed to make acids and alkalis fit into the pragmatic way in which we like to measure their effects. The use of a logarithmic scale in particular gives a tremendous spread around the half-way point on the graph, which is where our own measurements need precision.

So much for the background. What does this mean in principle?

– A very strong acid has a pH of 0 or 1.
– A very strong alkali has a pH of 13 or 14 (14 is the highest point on this scale).
– pH 7.0 is exactly neutral.

Blood, for example, is always slightly alkaline, with a pH of 7.35–7.4. Milk is slightly acid, with a rating around pH 6.5–pH 7.0. The tissue fluids of the human body must not be acidic in health, and fall in the range pH 7.1–pH 7.7. Outside that range, serious problems ensue.

What does this mean in practice? Well, some shampoos are made so that they have a stabilized (or 'buffered' as we say) pH that prevents the hair being subjected to an alkaline or acid condition. In normal use I cannot see that that is likely, but perhaps for delicate hair a buffered shampoo might be valuable. I would certainly never bother with it myself, for the hair is resilient and the slight stresses that are put on it by normal changes of pH are well within its capacity to survive. It seems almost like adjusting the central heating up or down by a fraction of a degree – fine in theory, pointless in practice.

One way in which the study of pH is valuable is that some organisms have a specific preference for the pH in which they can live. A frequent cause of bladder infections is the gut organism *E. coli* (p. 68) and it likes an acid environment. As it happens the pH of urine is typically on the acid side, around pH 5.0–6.5. If the pH goes above 7.0 then the *E. coli* organisms tend to die out. If you drink a solution of potassium citrate, known as mist. pot. cit., or the so-called 'citrate of magnesia' (which is actually made from citric acid, sodium hydrogen carbonate and magnesium sulphate) then the effect is to make the urine more alkaline. As the pH of the urine rises to 7.5 or more, there is no effect on the individual being treated, but the bacteria in the bladder cease growing and the cystitis they cause can be treated without any drugs at all.

More familiar to us than *E. coli* is the range of plants in our parks, gardens and farms, and they are often pH selective. Wheat, for example, can grow in alkaline soils at pH 8 that would be lethal to oats. Peanut plants need a strongly acid soil, less than pH 6, and so do the cone-bearing spruce, fir and pine trees, along with rhododendrons and raspberries. Holly can grow in soil at pH 4. A check on the pH of soil can be useful if you are having difficulty growing plants that are unsuited to your particular area. Soil treatments are commercially available to rectify an imbalance if that proves to be necessary. So there are general rules: animal tissues are typically on the alkaline side; plant juices are typically acidic; the sea is alkaline (around pH 8) and soils can range from pH 3 to pH 10. Just remember that below 7 is an acid, and above 7 is an alkali; then if someone tells you that the pH of a given wine is lower than it was, you may knowledgeably nod and agree that it should, then, taste a little more acid.

By the way, it isn't PH, ph, or even Ph; the H signifies the hydrogen ion, the p indicates the mathematical conjuring necessary to knock the values into shape, so pH is the only way to write it.

WHAT IS THE DIFFERENCE BETWEEN IRON AND STEEL?

Iron is a pure metal, an element with the symbol Fe (from the Latin for iron, *ferrum*). Steel is iron with additions. Iron usually contains a small percentage of carbon, which makes it into a harder metal; and in this form it is known as steel. Other components can be added to make the steel particularly easy to magnetise, or slow to corrode, or hard, or whatever, so that in this way a vast range of different types of steel is available.

Since iron is usually contaminated by a trace of carbon, almost all the iron we come across is actually steel when you get down to it. Pure iron is something that few people have ever seen. It is a white, bright-silvery metal which can be polished to a perfect mirror finish. It is soft enough to be drawn into wire, malleable so that it can be beaten into rods or sheets, and cannot be magnetised. Iron is an essential component of many body compounds, notably haemoglobin (p. 46), and a lack of it in the diet causes anaemia. In the form of ore, for it does not usually occur as the metal, iron is the fourth most common element in the earth's crust, amounting to 5.1 per cent. The occurrence of iron in meteorites may well have been the first source of the metal that primitive man tried to work, but the discovery of how to refine iron from its ore was so important for the development of the human species that it was named the iron age. Actually, it might have been better named the steel age, but no matter. . . .

HOW MANY KINDS OF STEEL ARE THERE?

There is no finite number. There are thousands of different kinds of steel for different purposes. What confuses the classification of steel is that the term 'iron' is used for one category of steel that contains a very high percentage of carbon, and therefore is the steeliest of them all. This is what we call cast iron, and it is nothing of the sort.

But let me explain. The purest 'steel' is, as we have seen, iron; it is a metal that rarely exists. Even the forms that are lowest in carbon contain 0.04 per cent carbon, and that is enough to designate it as steel. At the top end of the scale are the steels that contain 3.0 per cent carbon – that is *over seventy times as much* – and they are known as 'cast iron' once again. So, just as we have to accept that the 'iron age' was really steel, we must also realise that 'cast iron' is actually steel, too. . . .
The picture is further complicated by the other components found in steel, including naturally-occurring metals such as manganese, nickel and chromium, other elements such as phosphorus, sulphur and silicon, and metallic additives –

including chromium and vanadium – that are put in to provide specific properties.

What is interesting is that the traditional ways of producing steel – which used to involve stirring with green hazel twigs when the iron was melted, and urinating over the cooling castings – used to give a predictable quality that it takes high-tech to imitate today.

Here are the main categories of steel, and the purposes for which they are used:

Percentage Carbon:	Typical uses:
0.04–0.10	Low-carbon rimming steels, for making wire and – from the wire – nails etc.
0.10–0.25	Low-carbon structural steel, for making boilers, tubes, pipes.
0.25–0.45	Medium-carbon steels for forgings, steel castings etc.

0.45–0.75	Intermediate steels for rails, clock and watch springs (subsequently tempered) railway axles etc.
0.75–1.25	High-carbon steel used for piano wires, cutting tools, dies and ball bearings.
1.25–2.25	Cast iron for files, drawing dies, and finishing tools.
over 3.0	Normal cast iron, for piston rings etc.

WHY IS STAINLESS STEEL STAINLESS?

An English scientist, Harry Brearley, discovered in 1913 that if an alloy of steel with chromium (containing about 12 per cent chromium) was left exposed to moisture, it did not rust. This simple observation gave rise to the birth of stainless steel cutlery. There are now many different versions of stainless steel, not all of which are truly 'stainless', for some will discolour. They range from steels containing about 12 per cent chromium and less than 0.15 per cent carbon up to those containing 22–24 per cent chromium and 0.2 per cent carbon. Modern chrome-vanadium steels contain vanadium in addition, and many of the high-chrome alloys also contain up to 15 per cent nickel.

These components themselves resist corrosion, and in alloying them with iron in this way you can confer some of this characteristic on the steel that results. However it is not as obvious a benefit as it seems, for the blade of a stainless steel knife lacks the 'bite' of a normal carbon steel blade. For this reason, it is best to use stainless steel for table cutlery, where it looks good. Carbon steel knives remain best for the kitchen. They discolour, and they need to be kept dry if they are not to acquire rusty patches, but they can be honed against another blade or along a knife-sharpener, and the characteristic structure of the steel brings up a series of tiny tooth-like irregularities on the edge. This gives the blade a kind of microscopic sawing property, ideal for cutting in the kitchen, which a stainless blade cannot match.

HOW DO STEEL SHIPS EVER FLOAT?

Everyone wonders how this is possible. The answer lies in the principle of Archimedes (287–212 B.C.), who was born and lived in Syracuse, southern Sicily. Legend says that he had been asked by King Hieron to find a method of proving that a crown, supposedly made of pure gold, had actually been partly constructed from an alloy of gold and silver. Archimedes's answer was to weigh the crown in air, then in water, and measure the amount of water it displaced. Then he immersed equal weights of gold and silver, separately, in the same vessel of water, and noted the different amounts of water that overflowed. That gave him the answer. The buoyancy principle involved apparently occurred to him at a bath-house. He lowered himself into the bath-tub, filled to the brim, and realised that his own weight was related to the weight of the water that spilled over the side as he got in. This is the revelation, we are taught, that had him running back to his house in a state of excitement, shouting 'Eureka, eureka' (in Greek, $\varepsilon \upsilon \rho \eta \kappa \alpha$ = I have found [discovered] it) and – in his haste – naked from head to foot.

The principle states that a given body floating in water displaces its own weight of water, so that there is a force acting vertically upwards through the centre of gravity of the displacing body equal to the downward force due to the body's weight. Put more directly, a 1,000-tonne ship will displace a ship-shaped volume of water which will also weigh 1,000 tonnes. It is therefore possible to work out exactly how far down into the water a ship will sink at launch, for you will know the mass and the shape of the ship and from that can calculate exactly the corresponding amount of water that will be displaced by the vessel as it floats.

So long as the sides of the vessel are constructed so that they are taller than the depth to which the floating ship will come to equilibrium, then the ship will float. Modern vessels are marked with a line above which loading is not permitted, in order to ensure their safety at sea; but of course a water-tight ship fitted with hatches can theoretically survive any storm since it will float like any sealed container. Proposals have been put forward for the widespread use of submarine containers, in which the possibility of capsizing or foundering does not arise. This is how the popular London and New York landmarks, Cleopatra's Needles, were shipped from Egypt in the 1880s, so the proposition has adequate precedents.

WHAT ELSE CAN SHIPS BE MADE FROM?

Anything that is strong enough, and suitable for fabrication in the right shape, can make a ship. Many modern vessels are constructed from glass-fibre reinforced resins, and reinforced concrete has been used for medium-sized vessels. A concrete ship seems an oddity indeed, but so long as the principle of Archimedes is borne in mind (p. 81) then a properly-designed ship will always be able to float at a predetermined level.

IS THERE A SCIENTIFIC WAY FOR ANYONE TO LEARN HOW TO FLOAT?

Indeed there is. Swimming and sports manuals seem to concentrate on the need to relax, hold up the head, and so on; and many of them state that the human body is always lighter (i.e. less dense) than water so that floating is inevitable. That isn't true. A drowned body will often sink to the bottom of the sea, only to rise again when decomposition releases bubbles of gas in the corpse so that it bobs to the surface like a cork. In fact, the buoyancy of a human being depends on the amount of bone (which is dense) compared to the amount of fatty tissue (which floats). Many people trying to swim are actually denser than the water in which they are trying to do it, and so the idea that they are automatically destined to float is, simply, wrong. The idea could be dangerous.

The way to make sure that you do float, if you wish, is to rely on your lungs. Everyone is aware that water-wings will keep you afloat, for they exert an upward force on a swimmer that is equal to the weight of the water they displace (p. 81) and as they are pressed deeper into the water, the volume of displaced water goes up, and so too does the upward force of buoyancy. Water-wings have an important inbuilt feedback control in this way.

But your lungs can act as water-wings too. All you have to do, to become a first-time-in-your-life successful floater, is to keep your lungs full of air. Inhale deeply, hold your breath, and you are bound to float. That should give you a chance to feel your weight being borne by the water, to experiment with ways of stabilizing your position by making swimming movements with your hands or feet, and to gain confidence. The next stage is to do it for longer. Rather than hold your breath all the time you are floating, breathe out and in again quickly. Hold your breath again and carry on floating.

In this way you can survive on hasty changes of air in the lungs, by relying on this quick out-and-in technique of breathing. The lungs are fully inflated for 95 per cent of the time, and that keeps you afloat. Once you have gained confidence using this scientific method, then the rest of the swimming course comes that much more easily.

WHY DO DIVERS SUFFER FROM THE BENDS?

If a diver breathing air under pressures more than double atmospheric pressure comes to the surface too quickly, he is likely to suffer from the bends (also known as divers' palsy or caisson disease).

What happens is that the dissolved nitrogen in the bloodstream forms bubbles as the pressure is released. It is exactly like the effect of unscrewing a bottle of soda-water or pop. The gas (here carbon dioxide, CO_2) is held in solution under pressure, and when the pressure is released the gas fizzes out in the form of tiny bubbles. If bubbles of nitrogen forming in the blood lodge in the brain or in some other vital spot, then a diver can be permanently disabled. For this problem there are two answers:

(a) The diver must come to the surface slowly. This can either be done by raising him at predetermined intervals, a little at a time, so that he can exhale the excess nitrogen through his lungs in the normal fashion; or else he can be kept on board the recovery ship in a pressurised chamber that is slowly reduced to normal atmospheric levels.

(b) If the diver is given a mixture of helium and oxygen instead of nitrogen and oxygen, the lower tendency for helium to dissolve in the bloodstream greatly reduces the risk of the bends.

CAN ANYONE ELSE SUFFER FROM THE BENDS?

Yes, anyone who is subjected to a reduction of pressure is likely to throw out of the blood little bubbles of nitrogen. A diver who is coming to the surface too quickly is one example, clearly, but it is also a risk for a pilot of a plane or glider that is not fitted with a pressurised cockpit. A sudden ascent to high altitude can bring about such a drop in pressure that the blood releases nitrogen bubbles and the pilot suffers from the bends. Here the cure is much simpler than it is for the diver, for all the pilot has to do is descend quickly to a lower altitude and the problem will be immediately solved.

The answer is exactly the same as it is for the diver – i.e. increased pressure – but whereas a pilot is getting nearer the surface of the earth (and safety) as he descends, the diver is getting further away from the surface when he moves down to a lower level. For this reason the bends is much easier to treat if it occurs in a pilot; indeed he can usually cure himself before any long term difficulties have time to become established.

WHAT IS HELIUM?

Helium is a gas that is lighter than air. It is named after the sun (Greek $\eta\lambda\iota o\xi$ = sun) since it was discovered on the sun *before* it was ever known on the planet earth!

The original discovery was made by the great English astronomer Sir Joseph Lockyer (1836–1920), founder of the scientific journal *Nature*. In 1868 he discovered an unaccountable element in the spectrum of the vapours surrounding the sun. From the spectrum he could deduce something of the nature of this 'rogue element', but as it did not correspond with any known material on earth he called it helium and left it at that.

Not until 1895 was the element found in an accessible form. In that year it was isolated by Sir William Ramsay (1852–1916), who had already been active in the discovery of several other similar inert gases (in 1894 he and Lord Rayleigh discovered argon; neon, krypton and xenon were discovered in 1898). He detected it in 1895 in the mineral clevite. Helium has a very low melting-point, about 4°C above absolute zero, and it was not liquified until 1908. Helium occurs in the air, but only in the proportion of one part in 200,000. Its main source is in the gases produced by natural wells, the chief of which are in the State of Utah. The gas fields in this area yield 1.3–8.0 per cent pure helium, and they are all held as a US government reserve.

The reason it is used for divers is that it is far less soluble in the blood than nitrogen (as discussed on p. 83) and in addition it clears from the bloodstream 2.5 times more rapidly than nitrogen. However, the effect of the gas on speech is dramatic. Because of the change in the velocity of sound waves in helium, compared to their velocity in air, a person who takes a lungful of helium and then tries to talk normally will sound like Mickey Mouse instead: the voice sounds high in pitch and tonally flattened. For all the scientific interest in this gas, there are those two facts above all, that endear helium to me: it was discovered on the sun, 93 million miles away; and it makes you sound remarkably funny when you breathe it.

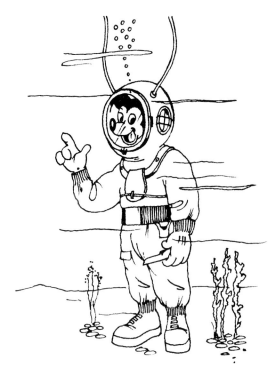

ARE HELIUM-FILLED AIRSHIPS
FEASIBLE, AND WHAT HAPPENED TO
THE GREAT AIRSHIPS OF AN EARLIER
AGE?

We have tended to dismiss airships as misbegotten leviathans of an ill-considered age, but at the time of their peak in popularity they were magnificently successful aircraft. As long ago as the period 1910–1914, when aeroplanes were tiny and unreliable, the German Delag Company was running a successful and accident-free transport network which carried a total of 10,000 passengers between German cities, and in 1929 the great airship *Graf Zeppelin* completed a graceful round the world tour, sponsored by the American newspaper tycoon William Randolph Hearst. No safe, luxurious aeroplane could compare with the easy passage that the airship achieved.

In the history of airships 173 rigid-framed airships took to the air of which 152 were built by the Germans. They were popular, too; the financial backing for the *Graf Zeppelin* was largely raised by Dr Hugo Eckener – chairman of the board – by public subscription. Britain's most

successful airship, the R 100, was designed by the veteran British designer Barnes Wallis through the Vickers company, also with speculators' backing.

However, these airships were filled with hydrogen, lighter than helium but dangerously inflammable. And this spelled the eventual end of the airship era. The first major disaster that heralded the change in opinion was the breaking-up of the R 38, which was built to American specifications by a British team. The designers based their work on crashed German airships recovered during the First World War, and guessed the loading factors that ought to have been incorporated as safety-margins. On a test flight with a mixed UK and US crew, the ship was run at full speed and then maximum rudder was applied. It broke in two and all but one of the Americans died.

In many ways this was regarded as a mishap, a chance of fate; and though it sowed seeds of doubt there were many people who were confident that safe and reliable ships could be built. One team at Vickers put forward proposals for what was to become the R 100. As luck would have it, the idea was under consideration when the first-ever Socialist government of Britain was elected. Socialism has nationalisation as a prime principle of belief, and this political ideal meant that the Government decided that, if airships were a good idea, then they were going to build one too.

There was much ill-feeling over the issue. Public attention focussed on the state-sponsored R 101 as the privately built R 100 simultaneously took shape. Part of the disquiet stemmed from the fact that the same ill-informed 'experts' who had designed the R 38 were now designing the Socialist airship, and some designers working on the rival R 100 stated publicly that they felt these men should have been charged with manslaughter because of the earlier disaster, rather than being entrusted with a second project of the same kind.

But the R 101 was a classical example of design by committee. It was made with stainless steel girders (p. 80) which proved difficult to work. The swept fins were of such a poor design that they became ineffective at small angles, which made the ship dangerously uncontrollable. The builders fitted heavy diesel engines, weighing 17 tons (as opposed to the R 100's 9 tons). An extra engine, amounting to 3 tons of extra lost payload, was carried for manoeuvring astern as the propellers were found to break if they were used in reverse pitch. The R 101, long on theory, short on fact, was fitted with costly and complex servo-mechanisms to help the helmsman steer, whilst the R 100 team had already calculated that the aerodynamic properties of the ship would make the controls light, and easy to manage without any such encumbrances . . . and so on.

In the end, the 'Socialist ship', as it was dubbed, turned out to have a lift of 35 tons, little more than half what was called for, which meant she did not have intercontinental capacity. The R 100, more inspired and more brilliantly designed by far, had 51 tons of useful lift and at 81 mph was 10 mph faster than the design had laid down.

Eventually a publicity trip was arranged for the R 101 to cruise to India. At the time she had never flown under full power, had never been subjected to an independent examination of engineering fitness, and had proved to be difficult to control in flight. Most important of all, she had no Airworthiness Certificate, but only a limited Permit to Fly over Britain for test purposes.

To a state department this is no obstacle: the Air Ministry simply wrote out their own Airworthiness Certificate, and the journey began. After seven hours of flight the R 101 had travelled only 200 miles, to Beauvais, north of Paris. It was then she met her end. A gas tank split, and the airship dived gently into the woods, erupting as she did so into a pillar of flame visible for miles. The effect of the accident in Britain was profound, for the R 100, though reliable on every test (and it even survived a cross-oceanic trip in which severe weather damaged her gas-tanks) was dismantled. It was the end of the British experience of hydrogen-filled airships.

That was on 5 October 1930. The end of America's dream came on 4 April 1933, when her own *Akron* – which had a maximum cruising speed of 47 mph – crashed into the sea off the New Jersey coast.

Finally came the dramatic end to the great German airship, the *Hindenburg* (or the LZ 129, as the Germans preferred to call her) on the evening of 6 May 1937. This huge craft, one-sixth of a mile in length and capable of prolonged cruising at 60 mph, was landing at Lakehurst, New Jersey, when it caught fire. The start of the disaster was, astonishingly, witnessed by two crew-men who were inside the hull checking the gas-bags when they heard a small explosion and saw a flash of light. Within seconds the highly inflammable gas-bags were exploding, the burning hydrogen reaching thousands of feet into the evening sky as a radio commentator was screaming into his microphone: 'It's burst into flames! It's burning, falling on the mooring mast and all those folks . . . this is one of the worst catastrophes in the world . . .'

Within half a minute it was over, and the great ship was nothing but a pile of twisted, red-hot girders. By a miracle, two of the survivors were the two crew-men who had been the only witnesses of the start of it all. Later investigations suggested that the ship had been sabotaged by an anti-Nazi fanatic, and some diligent investigative journalism has even located one person, Eric Spehl, who might have planted a small device in response to his Communist sympathies.

But whatever was the cause of the *Hindenburg* disaster, the end was now spelled out for airships. The three nations of Britain, the United States and Germany had all seen tragic ends befall their standard-bearing airships, and public opinion would have no more of it.

Since the Second World War there have been a few attempts to build airships filled with helium instead of hydrogen. Helium has twice the density of hydrogen, but is non-inflammable. The experimental airships have been used for experimental purposes, and for aerial television assignments (such as the Royal Wedding in London of 1982). Most of them have become familiar through the prominent advertising messages they carry from their sponsors. The most recent arrival on the scene was the *Skyship 500*, a helium-filled airship funded from commercial sources in the City of London. On 28 April 1982 it flew over Tower Bridge as a gesture of gratitude to the investors.

Many specialists claim that there is a sound commercial future for airships. Certainly, since they would be lifted by a gas that was not generally available in the 1920s and 1930s, the incendiary hazard that haunted the great airships of the past would be absent.

HOW DOES A JET WORK? WHO INVENTED THE JET ENGINE?

The jet engine is easy to understand. Most explanations tend to get ensnared in details of 'multi-stage axial compression' and 'concentric turbo-prop drive-shafts' but anyone who has blown on a dying fire to make the flames burst up again, or has blown onto a child's toy windmill to make the blades spin round has gone half-way to inventing the jet.

The principle lies in constructing two fans, or turbines, joined together on the same shaft like the wheels at each end of the axle of a model car. These turbines are mounted inside a tube which is just wide enough to allow them to rotate freely. Between the two sets of blades is a fuel inlet. When the jet engine is running, air is being sucked in by the first of the fans. It blows past the fuel intake and the effect of the rush of air on the burning fuel is to set up a roaring, fierce combustion. The hot gases passing on through the tube are now moving much faster than the incoming air, because of the expansion caused by the gases of the combustion process; so as the exhaust forces its way past the second fan – at the rear end of this simplified little jet engine – the effect is to blow these blades around faster still.

But of course, this makes the fan at the front turn faster, blowing in more air, which increases the rate of combustion, and in turn passes still more energy to the rear fan. Once the engine is running, you can control its speed by the amount of fuel that you allow into the fuel inlet.

The designers of jet engines have to face many technical problems which make actual jet turbine units more complicated than the basic two fans. The rate at which the turbines spin can cause vibration in the blades, and they can fail through fatigue, so the blades are made of special alloys, or metals like titanium, or they may be strengthened with carbon fibres, to stop this from happening.

Jet engines are tested so that they will not break up even if a bird gets sucked into the finely-balanced fans: the prospect of metal blades flying out at speed is one that has made the air licensing authorities very strict in their demands for jet engine safety.

There are several rows of fans in each turbine assembly, and the front turbine (the compressor) is usually made up of a profusion of carefully shaped blades that serve to compress the incoming air to maximum efficiency. The actual chamber in which the fuel is injected is specially tapered to generate the maximum amount of energy, and the rear turbine – which drives the compressor – has to withstand the temperatures and pressures of the exhaust jet.

In many modern jets the compressor is in two separate units, with a large set of fan-blades at the front of the engine to undertake a preliminary compression before the air is forced into the main compressor itself. These jets are more efficient, and they make less noise, than the single-stage compressor jets. The size of these huge fans can be so great (a diameter of more than 2 metres – tall enough for an adult to stand inside – is typical) that the ends of the blades are moving round faster than the speed of sound! This develops a characteristic noise, a kind of rumbling or roaring, which is less offensive to the ears than the scream of a typical jet. But the effect of this pressure loading on the blades is considerable, and the design problems they involve are immense.

Just as these engines are the height of sophistication for a simple principle, the ram-jet is at the other extreme – it could hardly be simpler, for the turbine fans are left out altogether. Ram jets cannot be used to launch a plane from the ground, because they rely on the onward rush of air to sustain the combustion process. This kind of engine is basically a tube that is tapered at both ends. The air rushing in through the front feeds the burning fuel with oxygen, and the exhaust gas that is forced out at the rear of the engine generates the thrust that drives it forward. Ram-jets are sometimes used for high-speed military aircraft, which are launched from a mother-plane or sometimes with their own conventional jets. The ram-jet can then be turned on when the plane is already moving fast enough to sustain its simple mode of operation.

There is an intermediate form, known as the pulse-jet. Here you have a much slower-moving jet than a ram-jet, so slow in fact that the burning gases would be as likely to force their way out of the front of the engine as emerge from the rear. So the front is closed by a kind of shutter looking like a hinged venetian window blind. When the jet is moving forward through the air the shutters are blown open, and in this way a charge of air passes straight into the combustion chamber. Immediately the fuel burns, and the resulting semi-explosion forces the shutters closed so that all the force is directed backwards in a pulse of energy (hence the name pulse-jet).

With the shutters closed, the burning fuel is rapidly extinguished, since there is no fresh air to sustain it. As soon as the pressure inside the combustion chamber falls, the shutters blow open again, fresh air is admitted, and the fuel bursts into flame once more. The shutters literally bang shut,

a further pulse of energy is emitted by the rear of the engine, and so movement proceeds. Like the ram-jet, this kind of engine can only function when it is already moving forward through the air. But because of the shutter principle, it doesn't have to be moving as fast as a ram-jet. The main practical application of the pulse-jet on a large scale was during the Second World War when the Germans fitted them to their V1 'flying bomb' pilotless aircraft. These were launched from a sloping ramp and were driven forwards by a steam-powered piston. The fuel was lit by a form of glow-plug (a coil of red-hot wire connected to a battery) and the pulses were established – at least in theory – as the weapon began to move along its launch ramp. The shutters soon fatigued in this kind of jet, for obvious reasons; so for a short-lived device like the V1 the pulse-jet was ideally suitable.

Pictures of V1 weapons show a small propellor at the front. That was nothing to do with the propulsion system. Instead, the spinning propellor, driven round by the slipstream, was connected to a distance-measuring system and after a pre-set flying distance the fuel was cut off. The pulse-jet was extinguished immediately, the plane stalled, and in this way it fell onto its target.

I have stood with people watching a jet plane take off who have said to me that it seems surprising that the exhaust gases (which you can often see against the background because of the 'rippling' effect of the hot gases on the refraction of light rays passing through them) could ever blow such a heavy object as an aircraft up into the air. The question of lift is a different matter, but the jet does not force the plane forward by 'blowing against the air'.

Instead, the plane is propelled by the reaction against the pressure of the combustion. If the jet produces, say, a tonne of force pressing against the combustion chamber then the force acting against the opposite sides of the chamber is exactly equal. But the force exerted against the front of the jet engine is *not* balanced by pressure against the rear, since the rear is open and the gases can escape. It is this pressure that drives the plane forward, not the 'pushing' of the jet exhaust against the atmosphere.

The invention of the jet is not as simple as the conventional tales appear. Jet engines were not first patented by Sir Frank Whittle, nor was his the first kind of jet to fly. The earliest recorded proposals for propulsion by a jet of any kind was the design of Hero of Alexandria, one of the most distinguished of the early Greek philosophers. There has been some controversy over his exact dates, but around A.D. 150 (the earliest accounts say 100 B.C.) he invented a spinning toy driven round by a jet of steam.

But the gas turbine idea was first patented by an English scientist, John Barber, in 1791. The first engine which could be regarded as the forerunner of today's aircraft jets was patented by a Frenchman named Guillaume in 1920. During the 1920s a development engineer at the British Government's Royal Aircraft Establishment at Farnborough, A.R. Griffith, worked on a practical design and by 1929 he had made a working compressor turbine unit which was highly efficient. At this time he was transferred to the Air Ministry Laboratories in London, where the facilities did not allow his research to continue.

During this time an apprentice at the RAF College named Frank Whittle became interested in turbines, and in 1929 he applied for a patent on an engine that could use the gas-turbine idea to propel an aircraft. He tried to interest commercial firms in the idea, but neither they nor the Air Ministry in London were convinced that it had any merit worth supporting. So in 1935 Whittle allowed his patent to lapse.

Meanwhile, in Germany, an aerodynamics student at the University of Göttingen, Hans von Ohain, was working in a parallel programme of research of his own and in 1934 had patented his own version of a jet engine. Two years later the aircraft company of Junkers began work on a jet engine as part of a development exercise on gas turbines for other purposes. Their first design for a jet engine dates from 1938.

Frank Whittle was meanwhile successful in finding backing from a firm of investment bankers, and a small company called Power Jets Ltd was established. Their first jet engine was built and tested in April 1937. It often broke down, it proved to be hard to control, and there were

The Germans were close to perfecting a range of propulsion devices

difficulties in finding the right construction materials that could resist the temperatures and pressures generated by the jet. For two years the work stumbled on with little encouragement and with barely enough money to support it. Even so, important developments were made. For example, although Whittle knew nothing about turbine design (and he was repeatedly told by internationally-renowned specialists in the field that his views were 'nonsense') he could see that the accepted ideas behind the design of turbine blades was unsound. His individual labours revolutionised our understanding of turbine design, in spite of the fierce opposition he encountered from the establishment.

But by early 1939 support was unenthusiastic, and rumours circulated that the limited amount of support which the Government had authorised was about to be withdrawn. At this time intelligence reports from Germany suggested that the Luftwaffe might be close to perfecting a whole range of jet propulsion devices. By 1939, the German research centres were actively developing ram-jets, pulse-jets and jet turbines. So in June of that year the British switched from hostility to outright support, and Dr David Pye (later knighted), the Government head of Scientific Research, asked to see the jet demonstrated. He went away satisfied, and promised a greatly increased level of Government support.

Just two months later, in August 1939, the first jet-plane in history flew. It was powered by Hans von Ohain's design. Over the next few years, muddled planning and contradictory advice delayed the German jet development programme and even though there were several companies and a number of individuals working on jet design (often without knowledge of each other's interest) it was not until late in 1944 that their first production jet fighter, the Me 262, went into service. In England, meanwhile, the first Gloster jet fighter took to the air in May 1941. But jet aircraft on both sides emerged too late to make much effect on the outcome of the war.

It is worth noting that the idea for jet engines arose simultaneously in two countries, amongst people who were unconnected with the aircraft companies and who were largely untrained in engine development. If the matter had been left to the specialists it would probably never have seen the light of day.

The British switched to outright support

WHY IS GIN AND TONIC MORE INTOXICATING THAN A STRAIGHT GIN?

It seems impossible: you take a gin, dilute it with soda water (in the tonic), and it turns out to be more intoxicating than pure gin would have been. Impossible? It seems so, certainly. The reason is that carbon dioxide (CO_2, p. 82) seems to act as a means of accelerating the absorption of alcohol into the bloodstream. So in this way the alcoholic content of a glass of gin and tonic (which, after all, still contains the same quantity of alcohol as a neat gin) is encouraged to move from the gut into the bloodstream (p. 25) and so it produces its intoxicating effects more rapidly than would otherwise occur.

This explains why it is that a whisky and soda is more intoxicating than a straight Scotch, and also why champagne is more potent than a still white wine that contains just as much alcohol.

THE END